D1480941

PRAISE FOR
THE SERENGETI RULES
ALSO BY SEAN B. CARROLL

"Carroll is both a distinguished scientist . . . and one of our great science writers. . . . *The Serengeti Rules* is a visionary book, which celebrates the new wisdom and the men and women who have brought the vision to pass."

— *THE GUARDIAN*

"A compelling read filled with big, bold ideas." — *NATURE*

"*The Serengeti Rules* is wholeheartedly recommended for its entertaining view of biology from an original perspective."

— CLIVE COOKSON, *FINANCIAL TIMES*

"In *The Serengeti Rules*, the author goes from E. coli to elephants to lay out the basic rules that shape so much of what's around us and inside us."

— BRIAN SWITEK, *WALL STREET JOURNAL*

"Sean B. Carroll's new book *The Serengeti Rules* is a passionate telling of the story of the precarious and hard-fought balance that is the very precondition of health—both at the level of individual organisms and at the level of ecosystems. . . . The book is informative, well-written, and persuasive. . . . *The Serengeti Rules* is an optimistic book."

— ALVA NOË, NPR.ORG'S 13.7

"[A] triumphant account of how physiology and ecology turned out to share some of the same mathematics."

—SIMON INGS, *NEW SCIENTIST*

"*The Serengeti Rules* is one of the best biology books for general readers I've ever encountered. It should be required reading for every college student, regardless of major."

—ANDREW H. KNOLL, HARVARD UNIVERSITY

"*The Serengeti Rules* is a superb journey of a book written by a scientist of the first rank. Unfolding seamlessly from molecule to ecosystem, it explains with authority and grace why modern biology is central not just to human life but to that of the planet itself."

—EDWARD O. WILSON, HARVARD UNIVERSITY

"Original, provocative, and beautifully crafted, Carroll's book provides a glimpse into the deeper laws of biology that govern the earth."

—SIDDHARTHA MUKHERJEE, AUTHOR OF
THE EMPEROR OF ALL MALADIES: A BIOGRAPHY OF CANCER

"This book offers hope that we can make a difference, that we can follow those rules, and that things can get better on our planet, our home. It is well written, meticulously researched, and easy to read. I also learned more about the serendipitous nature of scientific discovery. I thoroughly enjoyed this book and highly recommend it to both teachers and students."

—CHERYL HOLLINGER, *AMERICAN BIOLOGY TEACHER*

A SERIES OF FORTUNATE EVENTS

Sean B. Carroll

A SERIES OF FORTUNATE EVENTS

Chance and the Making of the Planet, Life, and You

PRINCETON UNIVERSITY PRESS
Princeton & Oxford

Requests for permission to reproduce material from this work should be sent to permissions@press.princeton.edu

Published by Princeton University Press
41 William Street, Princeton, New Jersey 08540
6 Oxford Street, Woodstock, Oxfordshire OX20 1TR

press.princeton.edu

Library of Congress Control Number 2020939474

British Library Cataloging-in-Publication Data is available

Editorial: Alison Kalett & Abigail Johnson
Production Editorial: Ali Parrington
Text & Jacket Design: Chris Ferrante
Production: Jacquie Poirier
Publicity: Sara Henning-Stout & Katie Lewis
Copyeditor: Erin Hartshorn

Jacket art and illustrations on pages iii, 1, 15, 33, 61, 81, 97, 127, 149, and 163 by Natalya Balnova

This book has been composed in Skolar and Block Pro

Printed on acid-free paper. ∞

Printed in the United States of America

10 9 8 7 6 5 4 3 2 1

For my brother Pete who prodded me
to write something like this for years:
I hope it doesn't suck.

We live as it were by chance,
and by chance we are governed.

—SENECA (4 BC–65 AD)

All persons, living and dead,
are purely coincidental.

—KURT VONNEGUT (1922–2007 AD)

CONTENTS

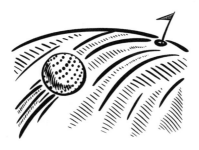

INTRODUCTION

THE TROUBLE WITH CHANCE

"When someone says everything happens for
a reason, I push them down the stairs and say,
'Do you know why I did that?'"

— STEPHEN COLBERT

PLAYING IN HIS FIRST professional tournament at the
Greater Milwaukee Open in 1996, Tiger Woods selected a 6-
iron from his bag on the tee of the 188-yard par-3 14th hole.
Although Woods was fifteen shots behind the tournament
leader, a large gallery had assembled to get a glimpse of the
heralded twenty-year-old phenom. Tiger launched the ball
into the wind, it landed about six feet from the pin, bounced
once to the left, and rolled straight into the hole. The crowd
whooped and whistled for several minutes.

It was not, however, the most auspicious start in the history of the game.

Comrade General Kim Jong-Il, playing the very first round in his life, was reported to have scored five holes-in-one at Pyongyang Golf Club in 1994, while en route to a 38 under-par score in which the then-future Supreme Leader of North Korea shot no worse than birdie (one under par) on any hole.

There are only two possible conclusions here: 1) Tiger is not such a big deal; or 2) somebody is lying. It is not hard for any of us, except perhaps North Koreans, to figure out which is the case.

If we were inclined to investigate further, we would discover that Tiger has recorded three aces in his 24-year career (a span in which he has won more than eighty tournaments). We would also learn that based on a large body of golf statistics, the odds of a professional golfer making a hole-in-one on any given par-3 hole is about 1 in 2,500. Tiger has played about 5,000 par-3s in his pro career so two aces would be expected; his career total of three is not extraordinary. However, the odds of an amateur golfer making a hole-in-one on a given hole is about 1 in 12,500; the odds of shooting two holes-in-one in the same round is about 1 in 26 million; and of sinking four holes-in-one is about 1 in 24 quadrillion (that's 24 followed by 15 zeros).

What makes Jong-Il's five aces even more amazing is the fact that, like most 18-hole golf courses, Pyongyang has only four short, par-3 holes. All other holes are at least 340 yards long. So, to get that fifth "1" in his round, the diminutive dictator must have been, in the immortal words of *Caddyshack*'s Carl Spackler (played by Bill Murray), "a big hitter."

We do not need any sophisticated understanding of probability, statistics, or the game of golf to doubt the veracity of the Dear Leader's scorecard. Nor, for that matter, do we have difficulty determining the improbability of the claim that young Jong-Il wrote 1,500 books and six operas during his three years at Kim Il Sung University. And what's the chance that, as was said, he really did not defecate?

. . . Even after that fifth hole in one!?

FALLING FOR FALLACIES

Puncturing fables about Kim Jong-Il (or his successors) is easy, but in other realms it pays to have some grasp of probabilities and the game, such as when our hard-earned money is on the line.

We flock to casinos in droves. About 30 million people visit Las Vegas each year to try their luck at various games of chance, including roulette, keno, craps, and baccarat, as well as slot machines. The house advantage in these games ranges from about 1 percent (craps) to 30 percent (keno). That's how the casinos can afford pyramids, gondola rides, shark tanks, fireworks, cheap buffets, and to pay Britney Spears $500,000 a night.

Nevertheless, we wager our hard-earned cash knowing full well that the odds are against our winning. Perhaps that is because even in these games of pure chance with dice, wheels, or electronics, most players believe or at least behave as if they can do something to improve their odds—by playing their "lucky" number, or betting on a "hot" shooter, or wagering on a color or number that is "due."

How does that work? Say, for instance, one is playing roulette and a black number has come up five times in a row. Should one keep betting on black, because black is "hot"? Or should one bet on red, figuring that a red number is "due"?

Does the bet change if black has come up ten times in a row? Or fifteen times in a row?

These questions are not hypothetical. On August 18, 1913, at the Casino de Monte Carlo, a remarkable run of black numbers unfolded at the roulette table. On European wheels, there are eighteen black numbers, eighteen red numbers, and one green "0," so a red or black number is expected to come up almost half the time. By the time black had come up fifteen times in a row, gamblers started placing larger and larger bets on red, convinced that the streak was due to end. And yet black hit again, and again. Players doubled and tripled their stakes, figuring that the chances were less than one in a million of a run of twenty consecutive black numbers. But the wheel kept hitting black until the streak ended at twenty-six. The Casino made a small fortune.

The incident in Monte Carlo is the textbook case for what has been dubbed the "Monte Carlo Fallacy" (or "Gambler's Fallacy")—the belief that when some event happens more or less frequently than expected over some period, then the opposite outcome will happen more frequently in the future. For *random* events such as rolls of dice or the spin of roulette wheels, this belief is false because each result is independent of the previous rolls or spins.

Our very powerful brains have trouble grasping this simple reality. If you think the incident in Monte Carlo is an isolated case from a less sophisticated, bygone age, consider the phenomenon that unfolded in Italy in 2004–2005. The Italian national SuperEnalotto worked at the time by selecting fifty numbers (from 1–90), five each from regional lotteries in ten cities. As more than a year passed without the number 53 being drawn in Venice, playing this *ritardatario* (delayed number) became a national obsession. Some citizens started betting so heavily that they exhausted family savings or ran up large debts. Despondent over her large losses, one woman drowned herself off Tuscany. A man near Florence shot his family and himself.

Finally, after almost two years, 152 draws, and more than 3.5 billion euros wagered on 53 alone (an average of more than 200 euros per family), the number was finally drawn in Venice, putting an end to what one group called the country's "collective psychosis."

Our problems with randomness in games spills over into real-life decisions. How many parents with children that are all of one sex opt to have another with the hope, if not the expectation, that the next child will be of the opposite sex? But, like the flip of a coin, the sex of a baby is pretty close to a random event. I say "close" because there is a slight skew in the natural birth ratio of boys to girls of about 51:49.

The Monte Carlo Fallacy is an example of what psychologists call cognitive bias—errors in thinking that skew the way we see the world. When gambling, these biases distort our sense of control over random outcomes and cause us to overestimate our chances of winning. A large body of research has revealed that our cognitive biases and our responses to them are part of our normal brain wiring. Psychological studies on both laboratory subjects and in real field situations (casinos) have documented the Monte Carlo/Gambler's Fallacy concerning runs of numbers. They have also found that near misses of jackpots (non-wins that fall close to winning combinations) increase our motivation to play.

One explanation for our fallacious thinking is that our brains are adapted to working every day to perceive patterns and to connect events. We rely on those perceived connections both to explain sequences of events and to predict the future. We can easily be tricked then to believe that some sequence is a meaningful pattern, when in fact a string of randomly determined independent events is just that—random.

It is a matter of our biology, then, that humans have such a complex relationship with random chance. On the one hand, we sure do enjoy games of chance, even though we lose often. Of course, when we lose, we accept it as just a matter of "bad luck."

But on the other hand, when we win—and many people do win every day—that often gets an altogether different interpretation. Good fortune is often chalked up not to the mathematics of chance, nor even to mistaken confidence in gambling "strategies," but rather to other forces. For some it is a just reward for good character or deeds, to others it is a prayer answered.

Take California truck driver Timothy McDaniel. On Saturday March 22, 2014, McDaniel lost his wife to a heart attack. The next day, he bought three "Lucky for Life" lottery tickets. When he scratched them off, he discovered he had won $650,000. McDaniel said, "I think she just kind of sent me this money so I could continue taking care of the (grand) kids."

McDaniel's heartbreaking story reflects how in the larger game of life and death our relationship with chance is even more conflicted. Many people prefer to banish chance altogether, to believe that, as McDaniel told reporters, "everything happens for a reason."

But not everyone.

THE PRINCE OF CHANCE

Jacques Monod grew up just down the coast from Monte Carlo in Cannes, France, another town famous for its casinos and, later, its film festival. Graced with movie star looks—one prominent French journalist described him as a "prince" who resembled Hollywood icon Henry Fonda, as well as considerable musical talent, and an exceptional intellect, Monod struggled to decide on a career path through his twenties. After distinguishing himself in the French Resistance, Monod rose to fame not as an actor or musician, but as a brilliant biologist. He shared the 1965 Nobel Prize in Physiology or Medicine for seminal discoveries about how genes work.

A pioneer in the field of molecular biology, Monod was privy to the blizzard of discoveries in the 1950s and early 1960s about the molecules that determined the characteristics of living things—what Monod and others dubbed "the secrets of life." He kept close company with a relatively small international community of leading researchers. For example, when James Watson and Francis Crick cracked the structure of DNA (deoxyribonucleic acid) in 1953, Monod was one of the first with whom Watson shared the breakthrough.

But as a Frenchman steeped in his culture's deep philosophical traditions, Monod was interested in science for more than just science's sake. After the war, Monod befriended France's leading philosopher-writer Albert Camus, and the two men pondered questions of human existence in Left Bank cafés. Monod felt that the public misunderstood the principal purpose of science as being the creation of technology. Rather, Monod believed technology was merely a by-product. He said, "the most important results of science have been to change the relationship of man to the universe, or the way he sees himself in the universe"—a relationship of equally intense interest to his friend Camus.

Monod thought that there were profound philosophical implications of the new molecular biology, particularly in the realm of heredity, which had gone largely unnoted in the broader culture. Several years after his Nobel Prize and Camus' untimely death, he decided to write a book to try to bring the meaning of modern biology to laypersons.

"[T]he 'secret of life' . . . has been laid bare," he wrote. "This, a considerable event, ought certainly to make itself strongly felt in contemporary thinking."

Monod used several chapters to describe the insights that had very recently emerged from the study of DNA and the deciphering of the genetic code. He understood this knowledge would be unfamiliar to most readers, so he included an appendix with chemical structures of proteins and nucleic acids, and a primer on how the genetic code worked. In a matter-of-fact style, he explained genetic mutations as accidental alterations—substitutions, additions, deletions, or rearrangements—in the text of DNA, in the sequence of the long strings of chemical bases (ACGTTCGATAA, etc.) that make up genes.

Then, almost without warning, he turned to the broader implications of how mutations arise in DNA. It is worth quoting him at length for after 111 pages of background, he delivered one of the most powerful ideas in five centuries of science (all *italics* are original):

"We call these events accidental; we say they are random occurrences. And since they constitute the *only* possible source of modifications in the genetic text, itself the *sole* repository of the organism's hereditary structure, it necessarily follows that chance *alone* is at the source of every innovation, of all creation in the biosphere.

"Pure chance, absolutely free but blind, at the very root of the stupendous edifice of evolution: this central concept of modern biology is no longer one among other possible or even conceivable hypotheses. It is today the *sole* conceivable hypothesis, the only one that squares with observed and tested fact. And nothing warrants the supposition—or the hope—that on this score our position is likely ever to be revised.

"There is no scientific concept, in any of the sciences, more destructive of anthropocentrism than this one."

In essence, heretofore obscure discoveries in biochemistry and genetics (largely studied at that time in simple bacteria) had upended two millennia of philosophy and religion that put humans at the center or apex of creation. "Man was the product of an incalculable number of fortuitous events," Monod wrote. "The result of a huge Monte Carlo game, where our number eventually did come out, when it might not well have appeared."

Le Hasard et la nécessité (Chance and Necessity) appeared in France in October 1970. It was a fairly technical book with several chapters on philosophy and genetics, and those appendices full of chemical diagrams. A first-time author, Monod did not know what reactions to expect.

The *merde* hit the fan.

The book received dozens of reviews across France and quickly became a bestseller—second only to the French translation of Erich Segal's *Love Story* (this was France after all). After it was translated into English, reviews and interviews with Monod were featured in several of the most prominent British and American newspapers and magazines.

Many commentators immediately recognized the threat chance posed to traditional ideas of humanity's origins and purpose. To Arthur Peacocke, a British biochemist turned prominent theologian, Monod had put forth "one of the strongest and most influential attacks of the century on theism." A flurry of articles and books appeared with titles such as *Anti-Chance: A Reply to Monod's Chance and Necessity*, *Beyond Chance and Necessity*, and *God, Chance, and Necessity*. Monod was invited to debate philosophers and theologians both in France and abroad, on television, radio, and in print.

American Calvinist theologian and pastor R.C. Sproul summed up the high stakes posed by chance in the first page of his book *Not By Chance*:

"It is not necessary for chance to rule in order to supplant God. Indeed, chance requires little authority at all if it is to depose God; all it needs to do the job is to exist. The mere existence of chance is enough to rip God from his cosmic throne. Chance does not need to rule; it does not need to be sovereign. If it exists as a mere impotent, humble servant, it leaves God not only out of date, but out of a job."

More than two hundred pages later, Sproul concluded: "Chance as a real force is a myth. It has no basis in reality and no place in scientific inquiry. For science and philosophy to continue to advance in knowledge, chance must be demythologized once and for all."

Sproul and other critics argued that what scientists perceived as chance merely reflected a lack of knowledge of true causes. Perhaps that was the expression of hope to which Monod alluded—the hope that as scientists learned more, our position on the role of chance would somehow be revised.

A SECOND CHANCE

The ensuing fifty years have not played out as either Monod or his detractors hoped. The Frenchman thought that the new insights from molecular biology should be a turning point for modern society—away from traditional beliefs about causes in the natural world toward one that embraced randomness and our chance existence.

Ha! Fat chance. The excitement and fuss stirred by *Chance and Necessity* simmered down, and Monod passed away a few years later. Surveys reveal that the majority of Americans, for example, continue to believe that everything on earth happens for His reasons.

But Monod's critics should take no comfort. The province of chance in the biosphere and human life has been revised, although not at all in the scope or direction that they hoped. The domain of chance has expanded into realms neither Monod nor anyone else imagined.

As we have learned much more about the history and workings of the planet, we have been startled to discover how the course of life has been buffeted by a variety of cosmological and geological accidents—without which we would not be here. As we have explored human history, we have seen how pandemics, droughts, and other civilization-changing episodes have been triggered by random, singular events in nature that easily might not have happened. And as we have probed human biology and the factors that impact our individual lives, we have caught chance red-handed, reigning over the often-thin line between life and death.

This book tells the stories that Monod could not—of astonishing discoveries from the planetary to the molecular scale, from great upheavals across the globe to the machinery of chance that operates within every cell of every creature, including ourselves. And while these discoveries vaporize the comforts of anthropocentrism, the story of chance, I hope you will come to agree, is much more than highfalutin philosophy or the refutation of theologians' wishful thinking.

I hope that you are awed—awed by the power and the drama of asteroids slamming into the planet, of continents colliding, and of the rapid rising and falling of ice and oceans; awed by the realization that we live on (and are at the mercy of) a planet that is far more unstable than our short lives perceive; awed by the knowledge of how random chance is at the source of all of the beautiful and wondrous creatures with whom we share the planet; awed by the unique invisible accidents that made each one of us; and awed by the fact that we humans, recent descendants of bands of hunter-gatherers who persevered through a period of exceptional chaos, have in just the last fifty years or so, figured all of this out!

My goal here is to be comprehensible without being comprehensive. It is almost trivial to claim that the world is the way it is or that we are here because of a long chain of chance, albeit fortunate, events. The explanatory power I seek comes from specificity. It is essential to unpack some of those events to appreciate how they shape the direction of life. The layout of the book follows a simple three-part logic. I'll begin with inanimate, external chance events that have shaped the conditions for life (Part One, "Stuff Happens"), and then turn to the internal random mechanism within every creature that generates the adaptations to those conditions (Part Two, "A World of Mistakes"). Then, I bring the story to the personal level (Part Three, "23 and You") and how chance impacts our natural lives, as well as our deaths. Our chance-driven existence shatters long-held beliefs about humanity's place and raises challenging questions about the meaning and purpose of our lives. In the Afterword, I'll offer some possible replies with the help of some special guests.

This is a relatively small book for a really big idea. Science has given us a handful of really big ideas over the centuries, but they have been received in funny ways. Darwin had a huge idea that was very simple to understand, and even though the evidence is massive and everywhere, many refuse to believe it. Einstein had a brand new idea, and even though few understand it or the evidence for it, most everyone seems to believe it. Monod had a great idea, but these days most people (other than scholars) have not heard of it, or of him.

My greatest hope, then, is that this short book might be chance's second chance.

PART I

STUFF HAPPENS

CHAPTER 1

THE MOTHER OF ALL ACCIDENTS

"The Age of Reptiles ended because it had gone on long
enough and it was all a mistake in the first place."

—WILL CUPPY, *HOW TO BECOME EXTINCT* (1941)

IN 2001, Seth MacFarlane was the twenty-seven-year-old
executive producer and creator of the not-yet-hit animated
show *Family Guy*. Having broken into the entertainment
big leagues at such a young age, MacFarlane was invited
back in September to address his alma mater, the Rhode
Island School of Design. After giving a talk, he went out for
what turned out to be a late night of drinking with some
professors.

The next morning, September 11, MacFarlane raced to
catch an 8:15 a.m. flight out of Boston back to Los Angeles. He was too late—the flight was actually at 7:45 a.m.;
his travel agent had written down the wrong information.

MacFarlane was rebooked on a later flight and went to doze in the passenger lounge. He was awakened by a commotion as passengers gasped at news coverage from New York showing the North Tower of the World Trade Center aflame. A short while later, the plane that struck the Tower was identified as American Airlines Flight 11 from Boston to Los Angeles—the flight MacFarlane had missed.

Actor Mark Wahlberg had also been booked on that same flight. A rising star known for his work in *The Perfect Storm* and *Boogie Nights*, Wahlberg and some friends changed their plans and hired a charter plane to a film festival in Toronto. They later flew on to Los Angeles.

Eleven years later, MacFarlane and Wahlberg teamed up to make the film *Ted*. So, just what are the odds that these two guys would both miss Flight 11, and later make a hit movie together? Were their escapes from mass murder just dumb luck, or was there a greater purpose at work? Were MacFarlane and Wahlberg spared so that our lives would be enriched by a pot-smoking, trash-talking teddy bear come to life? Or, so that movie industry coffers would be enriched by more than $500 million dollars?

MacFarlane himself doesn't think so. "Alcohol is our friend, that's the moral of that story," he offered. "I am not a fatalist."

Dumb luck, happenstance, accident—call it what you will. MacFarlane's late arrival to the airport was purely an accident, albeit an accident with enormous personal consequences. It is sobering to think what a thin line there can be between victim and survivor, between life and death. What a difference just 30 minutes can make.

It's a thin line in nature as well, not just for individual creatures (think of animal prey), or even species, but of whole worlds. Drive almost anywhere outside of a city, and the road is likely cut in places through a rock bed. Chances are most of us just ignore the pages of history staring us in the face. But those stacks of often colorful stone tablets tell stories, if you know how to read them.

Strada regionale (SR) 298 winds through a limestone gorge just outside Gubbio, a charming medieval town in the Umbria region of central Italy. In the mid-1970s, geologist Walter Alvarez saw an interesting pattern in a column of rock very close to the road (Figure 1.1). He noticed that in one section of the many layers of limestone, there was a switch in color, from white below to red above. When Alvarez looked closer, he saw that there was a peculiar layer of greyish clay separating the two colors of rock. Alvarez' decryption of that one centimeter thin line would lead to one of the most stunning and revolutionary scientific discoveries of the twentieth century and begin to tell the story of the most important day on Earth in the last 100 million years—a day that was very, very unlucky for most everything alive, but would eventually turn out to be extremely fortunate for us. And on that day a long, long time ago, 30 minutes would make all the difference.

A DIVIDING LINE BETWEEN TWO WORLDS

One way that geologists characterize rocks is by the fossils they contain. The Gubbio rock formation was once part of an ancient seabed, so it contained the fossilized shells of tiny creatures called foraminifera, or "forams" for short. These abundant, single-celled organisms are part the ocean's plankton community and food web. When forams die, their shells settle in ocean sediments and form parts of limestones. Different foram species, which have different sizes and shapes of shells, have existed throughout time and so can be used to assign rocks to particular time periods.

When Alvarez looked at the forams from the rock cut outside Gubbio, he saw that the white layer of rocks contained a diverse array of large fossil forams. But the reddish

FIGURE 1.1. The road cut at Gubbio, Italy

Right, Luis and Walter Alvarez at limestone outcrop, Walter (right) is touching the top of the Cretaceous limestone. © 2010 The Regents of the University of California, through the Lawrence Berkeley National Laboratory.

Bottom, the K-Pg boundary layer at Gubbio; the older white fossil-rich Cretaceous layer below is separated from the darker Paleogene layer above by a thin layer of clay that lacks fossils (marked by a coin). Photograph by Prof. Walter Alvarez/SCIENCE SOURCE.

layer of rock just above it lacked those species and contained only a few, much smaller species of forams (Figure 1.2). And the thin layer of clay in between the two colors of rock appeared to lack fossils altogether. Alvarez realized something dramatic had happened in the ocean that had driven many foram species extinct in a short period of time.

One thousand kilometers from Gubbio, at Caravaca on the southeast coast of Spain, Dutch geologist Jan Smit had noticed a similar pattern of changes in forams in rocks. Moreover, Smit understood that the line that lacked forams marked a well-known boundary in geology and Earth history, a dividing line between two worlds.

Below the boundary lay rocks of the Cretaceous Period, named for characteristic chalky deposits, and making up the last third of the Age of Reptiles when dinosaurs ruled the land, pterosaurs patrolled the skies, and mosasaurs preyed on ammonites (close relatives of the nautilus) in the seas. Above the boundary lay rocks of the Paleogene, which contain none of these creatures, but mark the beginning of the Age of Mammals, in which furry animals emerged to become the largest animals on land and in the seas.

The Cretaceous-Paleogene boundary (known as the K-Pg for short; formerly known as the K-T) marks not only the extinction of dinosaurs, pterosaurs, mosasaurs, and ammonites, but the mass extinction of about three-quarters of all species living around the globe 66 million years ago. Alvarez, Smit, and their colleagues wondered: What on earth could have caused the disappearance of widespread, tiny organisms like forams, as well as much larger creatures?

IT CAME FROM OUTER SPACE

The short answer, as you most likely have heard, is that it wasn't something on earth, but something from space.

Earliest Paleogene Forams ⎯⎯ 0.1 mm

Late Cretaceous Forams ⎯⎯ 0.1 mm

FIGURE 1.2. Foraminifera from the Paleogene (top) and Cretaceous (bottom)

Walter Alvarez and Jan Smit were intrigued by the rapid change in foram size and diversity between the end of the Cretaceous and the beginning of the Paleogene. Images courtesy of Smithsonian Institution. Photos by B. Huber.

But that short answer—the kind you see in newspaper headlines or textbooks—does not do the discovery or the event justice, nor tell us why this episode 66 million years ago means so much to understanding the role of chance in the world and our own story as a species.

Chemical analyses of the clay marking the boundary of the two periods, carried out by Alvarez, Smit, and their collaborators, revealed that it contained extraordinary levels of the element iridium, a material rare on earth but more abundant in certain kinds of asteroids.

The iridium in the boundary layer raised the possibility that the planet had been struck by an asteroid 66 million years ago, and that some dust from the spacerock had fallen over Italy and Spain. Before anyone got too carried away with such scenarios, it was important to look for signs of iridium at other K-Pg boundary locations. Sure enough, Alvarez found elevated iridium levels at a boundary exposed outside Copenhagen, Denmark, and another in New Zealand.

From the amount of iridium found in the boundary layer, Walter Alvarez' father Luis Alvarez, a Manhattan Project veteran and Nobel Prize-winning physicist, calculated the size of an asteroid that would be necessary to coat the globe in iridium (see Figure 1.1). He figured that the asteroid would have been about 6 miles (or 10 kilometers) wide.

That may not seem like a very big object compared with the diameter of the earth (8,000 miles, or 13,000 km). It is the same relative size as a BB pellet to a two-story house. But the key difference is that the asteroid travels much faster—about 50,000 miles per hour—so that upon entry into the atmosphere, the fireball's impact would be powerful enough to drill a crater 120 miles wide and 25 miles deep. That collision would blast huge volumes of debris and dust into and beyond the atmosphere, which would blot out the sun, quickly plunge the world into a deep cooling, and shut down plants' production of food.

Alvarez, Smit, and their collaborators forwarded the asteroid impact scenario for the mass extinction in 1980. It was a revolutionary, many would say radical, and some would say too radical, idea. Since the beginnings of modern geology in the early 1800s, the discipline had emphasized the gradual nature of changes on the planet, how slow but steady processes could create large changes over long periods of time. Geological science had replaced Biblical tales of floods and other catastrophes. The notion of a catastrophic event instantly rewriting the history of life was disturbing and too far-fetched for a good number of scientists to accept.

There was also the matter of the crater. One hundred twenty miles is a big hole, but no craters of that size or the right age were known. It was easy for critics to hold out in the absence of such evidence.

But in the following years, more exposed K-Pg boundary sites were discovered and studied (Figure 1.3). Several

FIGURE 1.3. The K-Pg boundary layer in southern Colorado

This is a terrestrial deposit that contains ejecta (white layer) from the Mexican impact site. Photo by Kirk Johnson, courtesy National Science Foundation.

around North America turned up some tantalizing clues. For example, large deposits of glass and clay spherules were found on Haiti. These small bead-like structures formed when molten rock that was blasted out of the crater quickly cooled as it fell back to earth. Shocked quartz grains that only form under the pressure of atomic explosions or impacts were also found on the island. Near the Brazos River in southern Texas, tell-tale signs of enormous tsunamis were discovered along with impact debris. This evidence pointed to an impact somewhere near the Gulf of Mexico.

Finally, in 1991, a 100-mile wide crater was identified that lies partly underneath the village of Chicxulub on Mexico's Yucatan Peninsula, and was shown to be of the very same age as the K-Pg boundary. The smoking hole had been found.

APOCALYPSE

Since the discovery of the Chicxulub crater, many kinds of scientists—geologists, paleontologists, ecologists, climatologists—have worked to unravel how the K-Pg impact triggered a mass extinction and to understand which species perished, which survived, and why? We now understand that the day of the collision, and the days, months, and years afterwards, were worse than the Alvarezes and Smit imagined.

The spacerock streaked across the atmosphere, covering the last 50,000 feet in one second. The collision induced earthquakes greater than magnitude 11 (100 times more powerful than the worst quakes in recorded history), caused the shelf of the Yucatan to collapse, and launched tsunamis more than 200 meters high that raced across the Gulf of Mexico and the Caribbean. The blast leveled everything on the landscape for 1,000 miles.

The enormous mass of rock blasted out of the crater was hurled in all directions: a thick curtain of ejecta traveling at several thousand miles per hour rained down across parts of North America, while the impact plume, consisting of superheated air, carbon dioxide, water and sulfur vapor, vaporized rock, and chunks of target rock, shot ejecta at velocities greater than Earth's escape velocity (about 25,000 miles per hour) into and beyond the atmosphere, which then fell back down toward the surface across the globe as trillions of red-hot meteors for hours.

The mass of this meteor shower was sufficient to coat every square meter of the planet with an average of 10 kilograms of spherules (areas near the blast received more, those farther away received progressively less). The immediate effect of this molten rain was to heat the temperature of the atmosphere to that of a broiling oven, estimated to reach 400–600 degrees Fahrenheit. The heat and falling debris were sufficient to ignite dry tinder and spark wildfires across the globe.

The fires in turn produced massive quantities of soot, which along with impact dust and enormous amounts of sulfur and water vapor were sufficient to dramatically reduce the sunlight reaching the earth's surface for several years, and to block photosynthesis and food production on land and in the ocean. Temperatures on land quickly plunged by 20 degrees Fahrenheit or more and were depressed for at least several decades. The massive injection of carbon dioxide into the atmosphere from the carbonate-rich impact site caused a rapid acidification of the oceans.

I know that sounds like the exaggerations of a Hollywood disaster movie, but this truly was an apocalypse.

The impact left its signs all over the globe. More than 300 K-Pg boundary sites have been identified across the planet. The effect on life is dramatically recorded in the fossil record across these boundaries. The world above the boundary (after the impact) is profoundly different from that below the boundary (before the impact).

The roll call of victims is much longer than the dinosaurs, marine reptiles, and ammonites. No non-aquatic animal much larger than a modern squirrel survived (squirrels themselves had not yet evolved). The reasons for this carnage are fairly clear: Life was baked, then frozen, then starved.

The immediate impact would have annihilated life only within 1,000 miles or so. The heat pulse, however, was global and would have made breathing very difficult for land animals everywhere. Those that survived the heat would have to endure the wildfires that destroyed forests and other vegetation that probably burned for weeks. And those that survived the fires would have to endure years of darkness and cold without new vegetation.

The extreme difficulty of life on land is vividly revealed by the plant fossil record. Far richer than the substantial animal record, paleobotanists can recover not only plant parts in rocks, but enormous quantities of spores and pollen, about ten thousand to one million pollen grains in a thimble full of sediment, that captures a rich picture of plant diversity at any given moment. The story told by pollen is one of utter devastation. In both hemispheres, there is rich and diverse pollen representing flowering plants and trees just below the K-Pg boundary and then at the boundary . . . almost nothing. Instead, just above the boundary layer there is a massive spike in fern spores. Unlike flowering plants whose pollen must land on a receptive flower to germinate, these ancient plants' spores are able to germinate where they land. Ferns are also the first plants to recolonize devastated habitats today, such as after volcanic eruptions. The fern spike persists for perhaps 1,000 years. The major plants that do eventually repopulate the world are very different from those that dominated it before the impact. Depending on the location, up to 78 percent of species went extinct.

Now that I have told you the story for the flowers and the trees, as Jewel Akens' old hit song goes, let me tell you about the birds and the bees. They got hammered.

Birds evolved in the late Jurassic (about 150 million years ago) after they split off from theropod dinosaurs (the same line that produced *Tyrannosaurus rex*). Many kinds of birds filled the late Cretaceous skies before the asteroid impact. After the global destruction of forests, most vanished. Similarly, there is evidence that bees, which originated in the mid-Cretaceous and evolved close mutual relationships with flowering plants, suffered a massive extinction after the impact.

Just think how bad conditions had to be to drive tens of thousands of plant species completely extinct. Now consider the fact that tiny foraminifera tell a similar story of conditions in the ocean, with 70 percent or more of planktonic species disappearing after the impact. Because plants and forams occupy the bases of land and marine food chains, creatures all the way up the chains collapsed with them.

But, despite all of the destruction, there were survivors.

HITTING THE RESET BUTTON

The earliest survivors have been found in the fossil record just above the impact layer. While each suffered large losses of species, all of the major groups of land vertebrates are represented among the survivors—reptiles, amphibians, birds, mammals.

The big scientific question is, why did certain species make it and others didn't?

Important clues have come from considering the lifestyles of those creatures that did make it. Crocodiles and turtles fared far better than their larger land-dwelling dinosaur cousins which were completely wiped out. Snakes as a group (not all species) also made it through the crisis. The few birds that made it appear to have been small, burrowing, ground-dwelling species, or shorebirds. The mammals that made it were also small, and probably burrowers as well.

Being aquatic or semiaquatic (crocs and turtles, shore-birds), then, appears to have been an advantage. So, too, with being a burrower (birds, mammals, snakes). This makes sense in terms of surviving the heat pulse. Small body size or slower metabolisms would also lower food requirements and would be an advantage in lean times. Small body size also affords a faster reproductive rate and more rapid recovery of populations.

As the landscape recovered, it would be up to these remaining bands of generally small creatures to fill the void, to refill the skies. Every species alive today is descended from these accidental pioneers.

This picture of the depleted ranks of surviving species repopulating the world has been confirmed by entirely different lines of evidence. Take birds, for example. We know that there are about 10,000 species of birds alive today. The fossil record gathered to date indicates that there were five major groups of birds in the late Cretaceous, four of which perished entirely. All modern birds come from the survivors of one group.

Most interestingly, we can get a good estimate of the timing of the origin of the forty or so major groups (parrots, falcons, hummingbirds, etc.) of modern birds by examining their DNA. For reasons I will explain later (Chapter 5), every species' DNA contains both a record of its ancestry and a record of the relative timing of its divergence from other species. A few years ago, a large group of researchers determined the entire DNA sequences (the genome) of representatives of all modern groups. That evolutionary "tree" of birds revealed that all living species descend from just a few lineages that survived the K-Pg mass extinction, and that bird evolution, pardon the pun, took off soon afterwards so that nearly all modern orders formed within about 15 million years.

The resulting pattern of bird evolution is like a menorah, with lines branching off of a common stem after the mass extinction (Figure 1.4). This pattern is paralleled by other animals. Frogs date back 200 million years, but the three major groups of frogs that include the vast majority of the roughly 6,800 modern species appear to have taken off after the mass extinction. The same is true of mammals. Both the fossil record and DNA evidence indicate that many, perhaps most, placental (i.e., not marsupial) mammalian orders (rodents, carnivores, ungulates, etc.) arose after, and some quite soon after, the K-Pg extinction event.

Think about how the great winnowing caused by the impact and its aftermath has shaped the direction of life. It would be like hitting a reset button and restarting the game of life with a few holdovers from the previous world. The great dinosaurs that had dominated the land for over 100 million years were gone. The world that came after, and its inhabitants, looked very little like the world before.

In terms of species, there are more birds than frogs, and more frogs than mammals. But we don't call the post-impact world the Age of Birds or the Age of Frogs, we call it the Age of Mammals. Part of that moniker reflects the fact that mammals soon evolved larger body sizes, and both vegetarians and carnivores came to occupy the space vacated by the larger dinosaurs. Mammals also took to the skies (as bats) and adapted repeatedly to water (as whales and dolphins; seals and walruses; manatees and dugongs). And of course, one group of mammals that arose after the mass extinction, the primates, eventually gave rise to us.

Which raises the question: Would we be here without the asteroid collision?

A LUCKY STRIKE

To answer that question, we need to weigh a few facts. First, mammals evolved well before the K-Pg extinction. They had coexisted alongside the great dinosaurs for 100

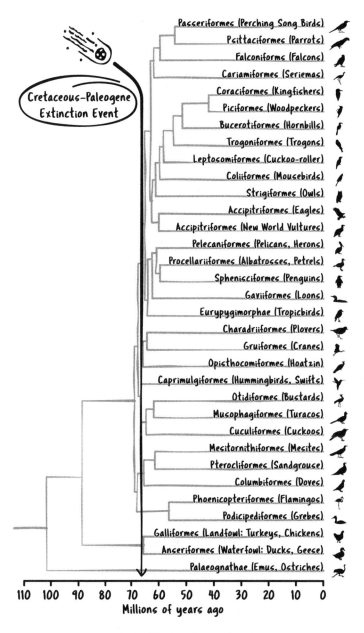

FIGURE 1.4. The menorah pattern of bird evolution after the K-Pg impact

There were several lines of birds that predate the impact; all but one vanished, while the surviving group rapidly radiated into the many forms we know today. Modified in part from Brusatte et al. (2015) with permission. Illustration by Kate Baldwin.

million years, and scores of species are known from various parts of the world from the late Cretaceous period. So, the presence of dinosaurs did not preclude furry mammals from arising. But second, mammals were relatively small-bodied, suggesting that they filled niches that the dominant dinosaurs did not. And third, within just a few hundred thousand years after dinosaurs disappeared, mammals became much larger than at any time in the previous 100 million years. This very rapid increase in average and maximum body size post-extinction suggests that the dinosaurs were a major force in limiting their size. It stands to reason, then, that without the asteroid impact the dinosaurs that had reigned for more than 100 million years would likely still be here, and therefore the primates would not be, and so neither would we.

The difference between the winners and losers was a matter of chance. The conditions triggered by the asteroid impact were beyond the experience of any creatures. Nothing in their evolutionary history prepared them specifically for years of hell. It was bad luck for the dinosaurs that the characteristics that made them dominant (e.g., large body size) made them vulnerable; good luck for a subset of mammals whose characteristics (e.g., small, burrowing) made survival possible, although not certain (the majority of mammals also perished).

The odds are not only against our existence without the asteroid, but the odds of an asteroid of sufficient size hitting the Earth are very low. The discovery of the K-Pg asteroid Chicxulub Crater spawned huge interest in other impacts. It turns out that no other asteroid strikes of the magnitude of the Chicxulub impact have occurred in the last 500 million years on earth or the moon (which receives a similar population of incoming bodies). To trigger a mass extinction, size matters. With an incidence of just one, all we can say is that Chicxulub is perhaps a 1 in 500 million years (or longer) event.

Moreover, it turns out that, even with a large asteroid, the location of the impact also matters. The rocks around the Yucatan target site are rich in hydrocarbons and sulfur, which resulted in the production of enormous quantities of soot and sunlight-deflecting aerosols. Geologists figure that as little as 1–13 percent of the earth's surface contain rocks that could have yielded a comparable stew of destructive materials.

This small target means that with the earth rotating at about 1,000 miles per hour, had the asteroid arrived just 30 minutes sooner, it would have landed in the Atlantic ocean; 30 minutes later, in the Pacific ocean. Just 30 minutes either way and the dinosaurs would probably be here, and there would be no *Ted* and, God forbid, no *Ted 2*.

CHAPTER 2

AN ORNERY BEAST

"Life's not about how hard of a hit you can give . . . it's about
how many you can take, and still keep moving forward."

—ROCKY BALBOA

ON OCTOBER 20, 1903, a crowd of more than 5,000 packed
into Philadelphia's Southern Athletic Club to watch local
favorite Joe Grim take on former world middleweight and
heavyweight champion Bob Fitzsimmons. Long before the
fictional Rocky Balboa, Grim had earned the love of the city
and the boxing world for his courage and pluck.

Born Saverio Giannone in Avellino, Italy, the eighth of
nine children, Grim came to America at age ten. He went to
work as a shoeshine boy and had a little stand outside the
Broadway Athletic Club. He loved watching the frequent
bare-knuckled boxing matches, so much so that one night
when a fighter did not show up and management asked for
a volunteer from the audience, Grim jumped at the chance.

Grim was pummeled, but remarkably bounced up like a
rubber ball after every knockdown. Lacking any technique,

Grim smiled and chuckled his way through the bout. He became an instant hit and soon gained a manager who hired him out to other athletic clubs and put him up against anybody and everybody. It was his manager who renamed him "Joe Grim" (nobody was going to try pronouncing his real name), and his fame spread as he took on opponents who just could not resist the urge to notch a sure win (Figure 2.1).

But they were in for a surprise. The five-foot-seven, 150-pound Grim fought bigger men and bigger names, including Jack O'Brien, Barbados Joe Walcott, Dixie Kid, Johnny Kilbane, and Battling Levinsky. They all knocked the hell out of him, but none could knock him out. Jack Blackburn, known for very quick hands and withering power, tried three times but could not flatten Grim. At the end of his bouts, the bloodied Grim would climb up on the ropes and shout to his adoring fans, "I am Joe Grim. I fear no man on earth," and say that he wanted to fight the world heavyweight champion.

He got his chance that October night with Fitzsimmons, a living legend who was said to be able to punch holes in anything. "The Freckled Wonder" had knocked out scores of great heavyweights on his way to two world titles, and is considered to this day one of the sport's hardest punchers. The bout was scheduled for six rounds; most fight experts expected that it would be much shorter. Fitzsimmons didn't think it would last one round.

Fitzsimmons knocked Grim to the floor again and again, only to see him "bob up serenely and assume an aggressive defense." Fitzsimmons just shook his head and kept swinging. Grim tried to protect his jaw as the champ bloodied his nose and ears. Fitzsimmons landed swings, hooks, jabs, and uppercuts that knocked Grim down four times in the third round and six times in the fourth. He stayed down seven times to the count of nine, only to bounce back up and resume the fight.

By the sixth round, the crowd started taunting the champ for not being able to put the Italian away. Fitzsimmons

FIGURE 2.1. Joe Grim

Photograph from *National Police Gazette*, No. 134, December 12, 1903, used with permission.

sprang angrily from his seat at the start of the round but Grim fought him off gamely, even landing a swing that momentarily stunned the former champ. Fitzsimmons flattened Grim again, landing him on his face. But Grim jumped back up and fought to the bell. Fitzsimmons promptly shook Grim's hand who, after seventeen knockdowns in the bout, made a celebratory somersault on his way back to his corner.

Grim had absorbed a battering "that would have killed an ordinary man," said one eyewitness. Grim's opponents were mystified how he could endure such punishment and still come up smiling. "I don't believe that man is made of flesh and blood," heavyweight champion Jack Johnson declared. Even though Grim lost almost all of his bouts and was knocked down hundreds of times in his career, his gritty performances brought him lasting fame as "The Human Punching Bag."

The moral of Joe Grim's story is that our species, or at least some of us, can take a punch. And that is a darn good thing because in the 66 million years between the time of the asteroid impact and when fearless Joe Grim arrived on the planet, the earth has thrown life a lot of punches. In the course of her tumultuous career, she has knocked out more than 99.9 percent of the species she has faced.

But not us, at least not yet.

So, what are the chances that Joe Grim and his fellow two-legged apes would have ever appeared? The odds against Joe's existence were a much longer shot than his very improbable run through that gauntlet of heavyweights. A constellation of planet-changing events has unfolded over the past 66 million years, some fairly slowly, some very quickly, any of which might not have occurred the way they did, when they did, or happened at all, and the story of life would be much different. Indeed, for the past million years, the earth has been locked in an exceptionally volatile cycle unlike any in the past 300 million years.

But whatever hasn't killed us has made us stronger. These upheavals have shaped our species' special abilities to roll with the punches that the planet throws at us, which is why we are still here and so many other contenders are not.

GOOD TIMES, BAD TIMES

The months and years following the K-Pg asteroid impact were without doubt some of the worst in earth's history. However, a trove of newly discovered fossils from Colorado reveal that even in one of the hardest-hit parts of the world, forests returned and mammals rebounded and evolved into new and larger forms within a few hundred thousand years. For many millions of years thereafter, life on earth enjoyed what may have been some of the best of times as mammals and birds branched out.

Geologists divide the last 66 million years into seven epochs of different lengths known as the Paleocene, Eocene, Oligocene, Miocene, Pliocene, Pleistocene, and Holocene (Figure 2.2). The boundaries between these epochs are typically marked by changes in rocks that reflect transitions in the conditions for life in the oceans and on land. While

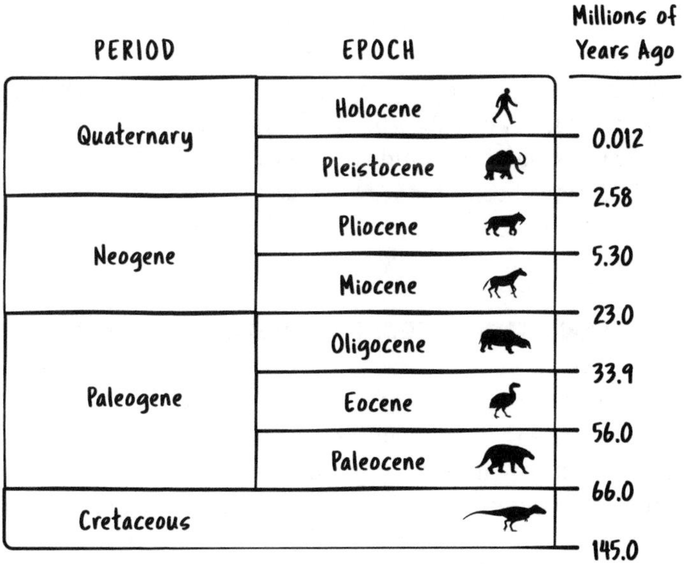

PERIOD	EPOCH	Millions of Years Ago
Quaternary	Holocene	
	Pleistocene	0.012
		2.58
Neogene	Pliocene	5.30
	Miocene	23.0
Paleogene	Oligocene	33.9
	Eocene	56.0
	Paleocene	66.0
Cretaceous		145.0

FIGURE 2.2. Geological Periods

Illustration by Kate Baldwin.

some boundaries mark significant extinctions, none are of the scope seen at the K-Pg boundary. Rather, they generally mark more limited turnovers of specific groups of plants and animals, including mammals, when certain forms may have emerged, disappeared, or their distribution changed on the globe.

For example, the transition between the Paleocene and Eocene is marked by a major extinction in deep sea forams. On land, however, mammal ranges expanded rapidly and the first primates appeared in North America, Asia, and Europe. The Eocene-Oligocene transition, in contrast, is marked by the extinction of more than 80 percent of placental mammals from parts of Europe and the disappearance of primates from North America. The end of the Pliocene brought the extinction of large marine animals including certain mammals, seabirds, turtles, and sharks, the latter of which included the infamous school bus-size *Carcharocles megalodon* shark. And the end of the Pleistocene just 11,700 years ago saw the extinction of most of the largest mammals (>100 pounds, or 44 kilograms) except those in Africa, encompassing some 90 genera that included giant ground sloths, camels, and saber-toothed cats in North America and the woolly mammoth and rhinoceros in Europe.

All kinds of scientists are drawn to these shake-ups and wonder: what happened?

CENE-CHANGERS: THE USUAL SUSPECTS

Many of the patterns across epoch boundaries have been known since the nineteenth century. The big question has been whether the turnovers in species were the steady coming and going of creatures over time, or the result of some abrupt event. Many candidates for the latter have been raised for virtually every change between "cenes" including asteroids, volcanism, supernovas, plate tectonics, falling sea levels, glaciation, or combinations thereof. The challenge has been to find tell-tale evidence for these events that is both contemporaneous with the cene-change and that can account for the changes to the planet and life.

Until fairly recently, the speeds of geological changes were not possible to pin down with great accuracy, nor was it possible to determine the magnitude of climatic change. So, the potential causes of these changes were difficult to sort out. One of the most powerful developments in geology has been the advent of techniques for peering back into past climates. Geologists can now infer ancient climatic conditions such as air and ocean temperatures by analyzing the relative abundance of stable isotopes of elements such as oxygen, carbon, and boron that are preserved in compounds found in the shells of forams or molluscs, or in buried sediments. These indirect chemical indicators, or "proxies," coupled with the development of highly accurate radiometric dating to determine the ages of rocks, have revolutionized our understanding of what transpired in the past and how quickly. While that power has not always led to a smoking gun concerning trigger mechanisms (yet), one crucial fact is clear: All transitions between epochs are marked by major, sometimes abrupt changes in climate.

For example, the remarkable speed of the Paleocene-Eocene transition was not appreciated until the paleoclimate record became accessible. It has been known for a century that several groups—even-toed ungulates (the group that would eventually include camels, deer, and others), odd-toed ungulates (the group that would include horses, rhinos, and others), and primates—make their first appearance at the beginning of the Eocene. The climate record revealed that the transition between the Paleocene and Eocene was accompanied by spikes in global deep sea and land temperatures of about 5 degrees Celsius (°C) and 5°C–8°C (9–14 degrees Fahrenheit), respectively, that lasted for about 100,000 years (Figure 2.3).

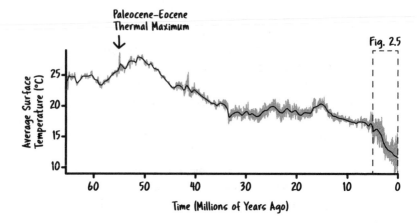

FIGURE 2.3. Global temperatures over the past 66 million years

The solid line is surface temperature averaged over 500 thousand-year intervals; the vertical hashes are the actual data points. Illustration by Kate
Baldwin based on Hansen et al. (2013).

Now you might think that 5 degrees is not such a big deal.
But consider that is a global average temperature change.
This swing would not be distributed evenly across the globe;
changes would be greater at higher latitudes than nearer to
the equator. To put that in perspective, a five degree drop
in surface temperature 20,000 years ago was sufficient to
bury a large part of North America, Europe, and Asia under
miles of ice. In the Paleocene, the effect was large regional
changes in climate, vegetation, and habitat.

Three later cene-changes are marked by cooling of surface temperatures: a relatively fast decrease of 4°C–6°C
from the late Eocene to early Oligocene, a gradual cooling
of about 3°C during the Pliocene into the early Pleistocene,
and a very sharp cooling and rewarming of several degrees
Celsius at the end of the Pleistocene immediately before the
Holocene. The challenge for geologists is to connect these
climatic changes to specific events.

Ever since the discovery of the Chicxulub asteroid impact as the trigger of the K-Pg mass extinction, asteroids have been at the top of the suspect list for other planetary changes. It is thought that spacerocks need to be 1–2 km in diameter to have a significant global effect. Impactors of this size will create craters 20 km or larger. The solar system has tossed a fair number of large rocks at the Earth over the last 66 million years; we know of a dozen such craters that meet these criteria:

CRATER	DIAMETER (km)	AGE (millions of years ago)
Boltysh	24	65
Chesapeake Bay	40	36
Chicxulub	150	66
Eltanin	N.A.	2.5
Haughton	23	23
Kamensk	25	49
Kara-Kul	52	5
Logancha	20	40
Mistastin	28	36
Montagnais	45	51
Popigai	90	36
Ries	24	15

Do any of these impacts (other than Chicxulub) match up with cene-changes? The answer: Not very well.

The Eltanin impact occurred in the South Pacific Ocean very close to the Pliocene-Pleistocene transition about 2.58 million years ago. The only known impact in a deep ocean basin, the asteroid is believed to have been about 2 km in diameter. The rock made one hell of a splash, blasting massive amounts of water and sulfur into the atmosphere and triggering megatsunamis. But it is not clear that would be sufficient to tip the planet into the long-term colder period that became the Ice Age.

Similarly, scientists have discovered impact debris right at the Paleocene-Eocene boundary off the coast of New Jersey. But that find alone does not prove the case. No crater is known (yet) of the exact same age so we don't know the size of the asteroid. Moreover, temperatures rose during the Paleocene-Eocene Thermal Maximum (PETM), while asteroid impacts are expected to cool the globe. Other events, such as massive volcanic eruptions or enormous releases of methane, are strong suspects for the warming.

Altogether, the evidence for impacts causing cene-changes during the Age of Mammals is scanty. There are no candidate impacts for most transitions, and there have been numerous large impacts without known lasting global effects (they certainly had regional or short-term effects). So, what else could explain rapid climatic swings and major shifts in plant and animal life? Recent sleuthing has revealed an altogether different kind of collision that transformed the world.

FROM THE HOTHOUSE TO THE ICEHOUSE

The map of the world today is not radically different than it was 66 million years ago in terms of the positions of most continents. But the climate is fundamentally different than that experienced by our early mammal and primate ancestors. From chemical proxies, we know that during the 15 million years or so following the asteroid impact, the planet was much warmer. Average global surface temperatures 51–53 million years ago reached 25°C–30°C (77–86°F), tropical forests attained their greatest ranges in earth's history, subtropical forest extended to the polar regions, and the planet was largely ice-free from pole-to-pole.

Today, the global average temperature is about 14°C (57°F), and ice extends across both poles. The planet has changed from the "hothouse" world of the early Eocene to an "icehouse" world. Paleoclimate records reveal that the average surface temperature has decreased in both gradual and more rapid phases, interspersed with some warming intervals (see Figure 2.3).

If not asteroids, what could explain this cooling trend?

Two key clues have been uncovered. The first is the timing of the glaciation of the Antarctic. Forty million years ago, Antarctica was in a similar position, but we know from fossils that it was a verdant landscape. In the late Eocene, however, it began to freeze over, and the massive continent was covered in ice by the early Oligocene and has been ever since. The glaciation of the Antarctic was a huge tipping point for the Earth's climate as it bound massive volumes of water in ice and lowered sea levels around the globe.

The second clue comes from the paleoclimate record of carbon dioxide (CO_2). Carbon dioxide is said to be the principal control knob on Earth's temperature because it is the major gas that helps trap heat in the atmosphere. CO_2 levels were extremely high in the balmy early Eocene, as much as 1,400 parts per million (ppm) (compared to 415 ppm today), but declined in the late Eocene and then fell precipitously to 600–700 ppm by the early Oligocene.

The drop in CO_2 would explain the major cooling across the Eocene-Oligocene transition and the onset of Antarctic glaciation. But what explains the big drop in CO_2?

There are several ways that CO_2 is drawn out of the atmosphere. Plants do it by converting CO_2 into food and biomass, some of which gets buried. The oceans do it by dissolving carbon dioxide in water. And rocks do it by chemical weathering: the carbon dioxide in rainwater forms carbonic acid that slowly dissolves rocks, which then release calcium, magnesium, and other ions into rivers and oceans, where they are combined with carbonate ions by shell-building creatures that eventually die and get buried. Could any of these mechanisms explain the decline and fall of carbon dioxide in the late Eocene? Was there more forest, more ocean, or more rock on the planet surface to do the job?

It turns out there was more rock, and for that we can thank India, or more specifically the Indian plate. This plate is one of the more than a dozen large, irregularly shaped tectonic plates that form a sort of global jigsaw puzzle of the world's continents and oceans. Made up of solid rock and spanning the Earth's crust and the upper mantle, the plates ride like rafts atop a semiliquid layer of magma and molten rock. Most meander fairly slowly, on the order of about 2–4 centimeters per year.

But the Indian plate is an exception. It was in a very different position 66 million years ago—in the *southern* hemisphere more than 4,000 kilometers south of the Asian continent and near Madagascar. Tectonic forces drove the plate northward at the unusually fast rate of about 18–20 centimeters per year until it collided with Asia about 40 million years ago (Figure 2.4).

Pioneering geochemist Wally Broecker has described that event as "the collision that changed the world." That slow-motion crash gradually built the Tibetan Plateau and the Himalaya mountains. And those rising mountains, by pulling more and more CO_2 out of the atmosphere in the late Eocene and since, set Earth's climate on a new path.

The Indian plate's faster movement appears to be just a fluke, a geological accident. The plate is about 100 km thinner than all of the other plates that were formed about 140 million years ago when a giant supercontinent called Gondwanaland broke into pieces. Its svelteness enabled tectonic forces to push and pull it along about 15 centimeters (6 inches) per year faster than other plates.

But what a difference 15 centimeters a year makes. That greater speed allowed the Indian plate to cover a much greater distance in 20 million years. At more typical speeds, the plate would not have yet hit Asia. The world's climate would certainly not have changed in the way it did, and the story of life would be much different.

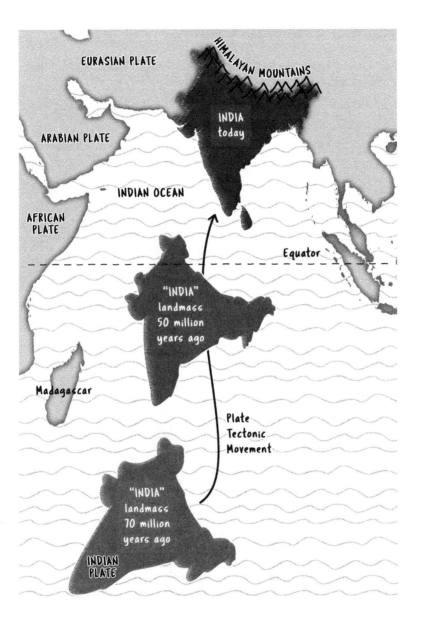

FIGURE 2.4. The collision that changed the world

The Indian tectonic plate moved rapidly north and collided with the Eurasian Plate about 40 to 50 million years ago, creating the Himalayan mountains and setting the planet's climate on a new, cooler path. Illustration by Kate Baldwin based on USGS (2015).

But India did smack into Asia, the planet changed, and it has continued to change in even more dramatic and surprising ways.

THE BIG CHILL

Look again at the temperature plot in Figure 2.3 above. Follow the bold, smooth line through time from 66 million years ago to today. The line is an average over 500,000-year intervals and shows the long-term cooling trend. Now, notice the spikier pattern superimposed on the main line; that plots the actual temperature data at many time points. Follow it down to the right and see how the spikes are getting taller?

To see those details better, here is the last 5 million years expanded:

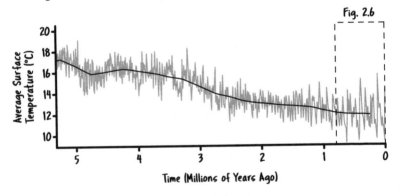

FIGURE 2.5. Global temperatures over the past 5 million years

The solid line is surface temperature averaged over 500 thousand-year intervals; the vertical hashes are the actual data points. Notice the hashes getting taller over the last few million years. Illustration by Kate Baldwin based on Hansen et al. (2013).

Look at the right end of the plot, the last 2 million years. Here is the last 800,000 years blown up in detail:

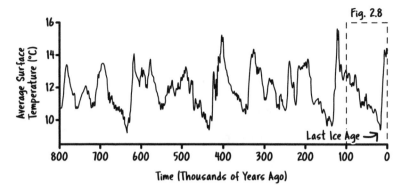

FIGURE 2.6. Global temperatures over the past 800,000 years

The declines in temperature are the glacial periods, when ice sheets advance; the peaks in temperature are the shorter interglacial periods when ice sheets retreated. Illustration by Kate Baldwin based on Hansen et al. (2013).

What was going on?

Welcome to the Ice Age. The long cooling trend since the Eocene has brought the Earth into its coldest period in 300 million years. That's right, *300 million* years. Those tall spikes indicate that for the past 2 million years the planet has been teetering up and down between two very different states. The temperature cycles reflect glacial and interglacial periods. The declines in temperature are the glacial periods when large ice sheets advanced across the Northern Hemisphere. In North America, for example, thick ice covered all of Canada, and glaciers reached as far as southern Ohio (Figure 2.7). The peaks in temperature are the shorter interglacial periods, when the ice sheets retreated. We are in an interglacial period now and have been for the past 11,700 years.

The trigger for the onset of the Ice Age is not certain. Whatever the catalyst(s), it is clear that CO_2 levels declined from a high of about 415 ppm in the Pliocene to about 280 ppm at the onset of Ice Age (Pleistocene). This new, lower level of CO_2 appears to be a critical threshold at which other

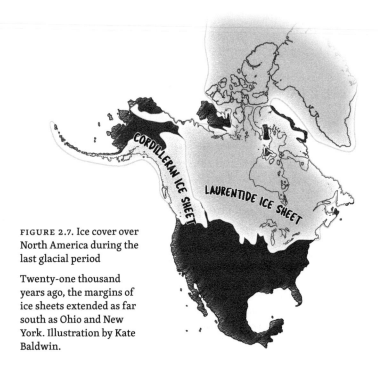

FIGURE 2.7. Ice cover over North America during the last glacial period

Twenty-one thousand years ago, the margins of ice sheets extended as far south as Ohio and New York. Illustration by Kate Baldwin.

mechanisms kick in (e.g., the reflectance of sunlight by ice) and drive further cooling.

But the cooldown is not static; it is cyclic. Something must then drive temperatures back up, before the planet cools once again. That something is the sun, or more accurately, the sunlight received by the Earth. The regularity of glacial cycles is driven by small variations in the orbit and tilt of the Earth that affect the amount of sunlight that hits the higher latitudes of the Northern Hemisphere. Over the last million years, the two most pronounced periodicities are about 100,000 years and 23,000 years long. The longer cycle is governed by deviations in Earth's nearly circular orbit caused by other large planets and determines the glacial/interglacial cycle. The shorter cycle is governed by the wobble in the earth's axis of rotation that is caused by the sun and moon and determines glacial/interglacial phases within longer glacial cycles.

The sun's radiation, however, is not the primary determinant of temperature. As the earth begins to warm, carbon dioxide is liberated from the oceans and increases the rate of warming. And as the ocean cools, it stores more CO_2. Over the past 800,000 years, CO_2 levels have oscillated between about 180 ppm and 280 parts per million (until very recently when it broke 400 ppm, but that is a different book!). The result is that atmospheric CO_2 levels and the earth's surface temperature are closely coupled over long cycles.

But long orbital cycles and carbon dioxide are not the entire story. It turns out that the climate during the Ice Age has been more tumultuous than anyone suspected.

FIBRILLATION

Let's zoom in below on a plot of Greenland temperatures (relative to today) over the last 100,000 years:

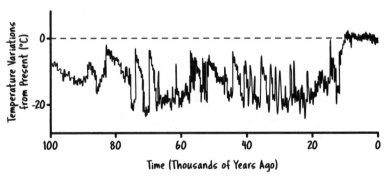

FIGURE 2.8. Rapid swings in Greenland surface temperatures over the past 100,000 years

Abrupt shifts of up to 20°C occurred in Greenland within glacial periods every few thousand years, which indicates an extremely unstable climate across the globe during this time. Illustration by Kate Baldwin based on data from ACIA (2004).

If you are thinking, holy #$%!, that's a lot of large swings up and down, you are thinking exactly what scientists thought when they discovered just how unstable Earth's climate has been—not only between glacial and interglacial times over the past 2 million years, but just within the last glacial period.

It was 1992 when rival European and American drilling teams reached three kilometers below the Greenland ice cap and discovered what they described as "violent" swings in climate. The ice, and trapped bubbles and dust within it, preserves a record of the climate when it was formed. Twenty-five times over the past 100,000 years, Greenland had gradually cooled and then warmed by as much as 7°C (12–13°F) *in just 10–20 years*. By comparison, the increasingly fast melting of Greenland ice that concerns scientists today has occurred with just 2–3°C of temperature increase over the past century. So, these 25 swings reflect great upheavals across the globe. What the heck was going on?

Here is one more graph; see if you can figure it out:

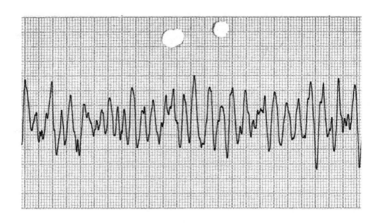

FIGURE 2.9. Ventricular Fibrillation

An electrocardiogram of a patient with an irregular heartbeat. Image from Beatson et al. (2011) by permission of Oxford University Press.

Stumped?

It is the electrocardiogram of a patient in ventricular fibrillation.

It is a pretty good analogy. Fibrillation is a form of arrhythmia—a chaotic, asynchronous, irregular beat where the muscle fibers contract randomly. We know now that, superimposed on the longer, regular rhythms of the orbitally driven glacial cycles, the earth's climate has been fibrillating for at least the last 800,000 years.

For example, about 12,900 years ago, as the earth was warming toward an interglacial state, some regions of the planet suddenly cooled by several degrees Celsius in a matter of a few decades. They then warmed again 1,200 years later in as few as five years. The period is known as the Younger Dryas for a diagnostic cold-adapted plant (Dryas octopetala) whose range expanded southward during the cold spell.

Why this or any other fibrillation occurred is still a matter of intense study. The timing between climate flipping events is random, so it is difficult to pinpoint any one factor. Rather, it appears that some combination of factors builds up to a tipping point, and the climate flips. One part of the explanation appears to involve the circulation of the Atlantic Ocean. The warm waters of the tropical Atlantic are carried north by the Gulf Stream, where they warm the atmosphere as the surface water evaporates. The surface waters then become saltier, colder, and denser, and sink in the subpolar region. This cold, salty deep water then flows southward past the equator, South America, Africa, and Antarctica and eventually returns to the South Atlantic and repeats the circuit. This Atlantic circulation acts as a heat pump that warms the Northern Hemisphere. But this pump gets flipped on and off—often.

The circulation and heat pump can be shut down by an influx of fresh water, such as from melting northern ice sheets, which can collapse very rapidly and result in rapid cooling. Alternatively, during colder times, there is less fresh meltwater, salinity increases, the circulation strengthens, and it can produce rapid warming.

Wally Broecker, who discovered the Atlantic circulation, was struck by how such mechanisms amplify climate change, rather than buffer against it. Shortly after the discovery of these rapid fibrillations Broecker remarked, "The paleoclimate record shouts out to us that, far from being self-stabilizing, the Earth's climate system is an ornery beast which overreacts to even small nudges."

Of course, the most interesting questions about this ornery beast concern the impacts of those overreactions on life. The asteroid took out most animals except those that spent most of their time in the water or below ground. What sort of animals could adapt to the irregular, abrupt, large swings in climate over the past 2 million years?

Put down this book and go look in the mirror.

You're looking at 'em.

AN ANIMAL FOR ALL SEASONS

To get an idea of what climatic fibrillations might mean to our species (and our ancestors), paleo-detectives have to turn their attention from the records entombed in Greenland's and Antarctica's ice to other kinds of clues buried in Africa's rocks and soil. Nearer to the equator, the story is not so much about cold and warm spells, but of wet and dry cycles. Dust, pollen, and lake and offshore sediments, for example, reveal intense fluctuations between moist and arid conditions.

To get some idea of the magnitude and pace of those wet and dry fluctuations, consider that while today the Sahara is the largest warm desert in the world (3.6 million square miles), from 5,000 to 11,000 years ago it was green! Beginning after the end of the last glacial period, the hyperarid

Sahara received abundant rainfall—more than ten times as much as today—that was sufficient to establish permanent lakes, and to support diverse plants, animals, and humans. Thousands of rock art paintings and engravings across current-day Algeria, Chad, Libya, Sudan, and Egypt record the elephants, hippos, giraffe, antelopes, and hunters of the "Green Sahara" before—in a matter of just a few centuries— the climate flipped back to an arid state and started making today's desert.

In order to understand the long-term consequences of the fluctuating climate on our and other species, paleo-detectives have focused on regions where there is a long record of the presence of humans and our ancestors. One such place lies forty miles southwest of Nairobi in southern Kenya, on the floor of the Rift Valley between two extinct volcanoes, Mt. Olorgesailie and Oldonyo Esakut.

Pioneering paleoanthropologists Mary and Louis Leakey first explored the eroding hills and gullies of the Olorgesailie Basin on Easter weekend 1942. They spread out to comb the white sediments and at almost the exact same moment, they called out to one another. Mary kept shouting at Louis to hurry over to see what she had found. "When I saw her site I could scarcely believe my eyes. In an area of fifty by sixty feet there were literally hundreds upon hundreds of perfect, very large hand axes and cleavers," Louis later recalled. Mary thought the scene looked as though a stone tool factory had only just been abandoned, although it turned out to be 700,000 years old. The scene was so startling and impressive that they decided to leave a large section exactly as they found it. A catwalk was built, the site was opened as a public museum, and Olorgesailie remains so today.

It also remains an area of intense research because the sediments there preserve most of the last 1.2 million years of history. For several decades, a large interdisciplinary team of researchers associated with the Smithsonian Institution in the United States and the National Museum of Kenya have labored to unearth that history. In addition to stone tools, animal fossils are also abundant, and together with climate proxies, they tell a dramatic story.

At least sixteen major environmental swings occurred from 1.2 million to 400,000 years ago during which the landscape fluctuated between wetlands and dry grasslands. The pace of change further accelerated over the past 320 thousand years. Yet, after most of these swings, tools are found in the subsequent layer of sediment, indicating that hominids were able to persist or at least re-colonize the area throughout a million years of highly volatile climate.

The animal bones, however, tell another story. For example, of the thirty mammalian species present about 500,000 years ago, including many large-bodied grazers related to giraffes, antelope, zebra, and elephants, only seven persisted over the next 200,000 years. These species were succeeded by sixteen new species not found previously at Olorgesailie. The simplest explanation for this large turnover in species is that the whipsawing climate brought about new conditions faster than species could adapt biologically.

What might explain the resilience of hominids? The stone tool record reveals some powerful clues. Over this same 200,000–year interval, the hominid tool kit underwent a dramatic redesign. Prior to 500,000 years ago, tools at Olorgesailie were predominantly large hand axes made from rock found in the basin. But between 500 and 320 thousand years ago, early humans started making more sophisticated tools, including points that could be attached to weapons and fine scrapers and awls (Figure 2.10). Many of these newer tools were fashioned from obsidian, a volcanic rock found as far as 50 miles away. In addition, these later toolmakers used color pigments, probably for body paint. The distant source of materials and new designs for these implements suggest that their makers had greater cognitive abilities and were engaged in more complex social behaviors than earlier residents of Olorgesailie.

FIGURE 2.10. Tools from Olorgesailie, Kenya

During a period of rapid climatic change, the tools used around Olorgesailie changed from large hand axes (left) to smaller, more finely crafted blades and points (right). Courtesy of Human Origins Program, Smithsonian Institution.

These were not their only skills. Hominids gained control of fire 800,000 to 1 million years ago and used it for hunting, cooking (more usable calories are obtained from cooked foods), and warmth. All of this knowledge would make these hunter-gatherers better able than other animals to roll with the punches of a fibrillating climate and less predictable resources (water, plants, game, and materiel).

And how might hunter-gatherers have gained greater cognitive and social abilities?

Bigger and better brains.

The hallmark of human evolution, hominid brain size has expanded dramatically, about three-fold, since near the onset of the Ice Age (Figure 2.11). Paleoanthropologists believe that there is a causal link between this unusually variable period of Earth history and the evolution of unusually large-brained, toolmaking animals capable of modifying and constructing their habitats.

We were born, then, out of a rare Ice Age set in motion by a long-ago geological accident and forged by one of the most

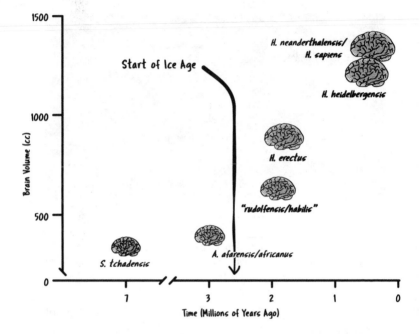

FIGURE 2.11 Human brain size increased dramatically during the Ice Age

After a long period of relative stasis, human brain size increased more than three-fold over the past 3 million years. Illustration by Kate Baldwin based on Bolhuis et al. (2014).

unstable and unpredictable climate cycles any mammal has encountered—a series of fortunate geological events. Today, thanks to our hardy Ice Age hunter-gatherer ancestors, we get to use our brains for more than just hunting and gathering, but for recreational pursuits—gardening, painting, writing books, etc.—or as punching bags.

PART II

A WORLD OF MISTAKES

WE HAVE PLAYED a bit of a game of "What if?" thus far. What if the cosmos had not thrown a fastball at the planet? What if the K-Pg asteroid missed the Yucatan? What if the Indian subcontinent inched more slowly toward Asia? What if the climate didn't whipsaw up and down during the Ice Ages?

The answers to all of these what-ifs in the inanimate, physical world are that the course of life would surely be so different that we would not be here, and so the explicit or implicit conclusion is that we are here by chance. I am about to double down on that position by digging into the role of chance within living creatures, but before I do, it is important to unpack just what I mean by chance.

Thus far, I have intentionally been vague about defining chance. In explaining the cause(s) of some particular outcome, scientists and historians recognize two related but distinct ideas—chance and contingency. The two terms do not have strict definitions in either discipline, and their colloquial usages vary. However, it is both useful and important to tease apart chance and contingency so I will define here what I mean when using each term.

By chance, I mean a rare, unpredictable, or random event, or one that entails so many variables or forces that is near enough to be random. The kinds of chance events we have seen so far include asteroid impacts; the formation, movement, and collision of tectonic plates; and the rapid flipping of the global climate.

Contingency I will use in its historical sense—a past event or process that was necessary for a particular outcome. If an outcome is dependent (contingent) upon a sequence of events or processes, such that it would not have occurred without each event transpiring as they did, then each event or step is a contingency.

Chance here, then, pertains to an event itself, while a contingency emerges through the benefit of hindsight. The interrelationship between the two is that a chance event can become a historical contingency through its effects. Contingency is the aftermath of chance. For example, we meet our spouses by chance, but that event becomes a contingency for the existence of our children. Similarly, the asteroid impact in the Yucatan was a chance event, but it was a contingency for the subsequent rise of mammals, primates, and our species.

The distinction is useful because it is pretty trivial and not very enlightening to say that the world or our lives are the way they are because of a sequence of contingencies. The central focus here is on chance, and the explanatory power I am seeking comes from *specificity*—from understanding the origin and significance of specific chance events that become those contingencies.

For centuries, the prevailing view of life was that nothing was left to chance or contingency, that all of its intricate complexity and beauty was perfectly designed by God in its present form, and unchanging. Indeed, that is what virtually all serious scientists once believed. In the next three chapters, we are going to peer deep into the machinery of chance that operates within every living creature. It is this internal, random process that generates the characteristics that determine whether and how creatures cope with the conditions of the external physical world. The living world, then, is where two entirely independent, chance-driven processes (external-inanimate and internal-living) meet. It is their intersection that produces all of the diverse forms that populate the planet.

CHAPTER 3

GOOD HEAVENS WHAT ANIMAL CAN SUCK IT?

"O MOST powerful and glorious Lord God, at whose command the winds blow, and lift up the waves of the sea, and who stillest the rage thereof; We, thy creatures, but miserable sinners, do in this our great distress cry unto thee for help; Save, Lord, or else we perish. We confess, when we have been safe, and seen all things quiet about us, we have forgotten thee our God, and refused to hearken to the still voice of thy word, and to obey thy commandments: But now we see how terrible thou art in all thy works of wonder; the great God to be feared above all . . ."

IT MAY SEEM like a verse straight out of a Monty Python skit, but this "Prayer to be Used in Storms at Sea" is from the canonical *Anglican Book of Common Prayer* (1789). And British naval captains and crews had damn good reasons to recite it. They also had many opportunities to pray because, with no science of weather forecasting, and many waters uncharted, nearly 1,000 Royal Navy ships were lost between 1700 and 1850.

Shipwreck was but one risk. Mutiny was another. It was difficult to maintain morale and discipline throughout long voyages as storms, hard labor, cramped living space, deprivations, drunkenness, and homesickness led sailors to revolt. The most famous mutiny, of course, was that on the *HMS Bounty* in 1789 when after five idyllic months of rest and free love on Tahiti, several sailors resisted leaving by wresting control of the ship from Captain William Bligh and setting him and eighteen others adrift in an open boat. In one of the greatest feats of leadership, navigation, and endurance, Bligh maneuvered 4,000 miles through miserable weather to safety on rations of an ounce of bread and a quarter-pint of water a day. It is probably not a coincidence, however, that Bligh was the target of two more mutinies in his career.

And if storms or mutiny did not sink a captain, there was always the specter of insanity. It was that very fear of a mental breakdown that led to young Charles Darwin's voyage on the *HMS Beagle*.

The ship's first Captain, Pringle Stokes, became utterly despondent two years into the *Beagle's* maiden voyage to South America (1826–1828). Anchored off the cheerily named Port Famine on the far southern coast of Chile, Stokes wrote in his journal:

"Nothing could be more dreary than the scene around us. The lofty, bleak, and barren heights that surround the inhospitable shores of this inlet were covered . . . with dense clouds, upon which the fierce squalls that assailed us beat . . . as if to complete the dreariness and desolation of the scene, even the birds seemed to shun its neighborhood . . . The weather was that in which . . . 'the soul of man dies in him.'"

Stokes shot himself a month later. Command of the ship was turned over to a lieutenant from another ship, Robert FitzRoy, who brought the *Beagle* home.

Three years later, in 1831, a second voyage was commissioned with FitzRoy at the helm. Not wanting to repeat the Stokes tragedy and concerned about a history of depression and suicide in his own family, FitzRoy sought to offset the loneliness and isolation he was certain to feel at sea. He told the Admiralty that he would like a well-educated gentleman to accompany him on what was to be a two-year voyage, someone who would share his scientific tastes, as well as meals and conversation. Darwin was not the first to whom the position was offered. Only after two others turned it down was Darwin—a 22-year-old "full of zeal and enterprize"— recommended for the post. Darwin put off his plans to study divinity and leaped at the chance to see the world.

The captain and Darwin got along very well, and the young naturalist was grateful for FitzRoy's seamanship. After rounding Cape Horn, the *Beagle* was pounded by fierce storms and almost capsized by a set of three huge waves but for FitzRoy's skill and command. Darwin wrote in his diary, "None but those who have tried it, know the miseries of a really heavy gale of wind. May Providence keep the Beagle out of them."

But after almost three years at sea, FitzRoy cracked. In late October 1834, when the *Beagle* was on the coast of Chile, and while Darwin was recuperating inland from an illness, FitzRoy resigned command of the ship and turned it over to a lieutenant. FitzRoy was irate that the Admiralty had admonished him for buying a supplemental surveying ship and refused his request for additional crew. Darwin explained FitzRoy's state in a letter home to his sister Catherine:

"We have had some strange proceedings on Board the Beagle . . . Capt FitzRoy has for the last two months, been working extremely hard . . . the cold manner of the Admiralty . . . & a thousand other &c &c has made him thin & unwell. This was accompanied by a morbid depression of spirits, & loss of all decisions & resolution. The Captain was afraid that his mind was becoming deranged."

The Admiralty's orders to FitzRoy in the event of such a development were made explicit in advance of the voyage:

"In the event of any unfortunate accident happening to yourself, the officer on whom the command of the Beagle may in consequence devolve, is hereby required and directed to complete, as far as in him lies, that part of the survey on which the vessel may then be engaged, but not to proceed to a next step in the voyage; as, for instance, if at that time carrying on the coast survey on the western side of South America, he is not to cross the Pacific, but to return to England by Rio de Janeiro and the Atlantic."

The *Beagle* was in fact on the western side of South America. The remainder of the voyage around the world was to be aborted. Darwin, however, could not bear giving up his adventures at that point. He concocted a plan to go off on his own to explore Chile, then venture overland to Peru, cross over the mountains into Argentina, and finally find another ship back to England. In either case, neither he nor the *Beagle* would see the Galápagos, or Tahiti, or Australia, or South Africa.

But FitzRoy regained his bearings and retook command. So, Darwin saw them all. And what he saw would change his thinking, and ours, about how the living world came to be.

It was Darwin, and a menagerie of wild and domesticated creatures, who first led us from a world created and governed watchfully by Providence to one operating entirely through natural laws. It was also Darwin who first replaced Providence with chance.

THE MYSTERY OF MYSTERIES

The central mystery in Darwin's day was the origin of species. The prevailing view, held by almost all scientists and laypersons alike, was that species were specially created by God in their present form and place, perfectly suited to their surroundings, and unchanging. Darwin, too, believed in special creation when he boarded the *Beagle* and had no inkling of any other idea for most of the voyage. But several encounters planted some seeds that would later germinate.

After FitzRoy's recovery, the *Beagle* and Darwin voyaged to the Galápagos Islands in September 1835. A prodigious collector of everything—he would amass more than 5,400 plant, animal, and fossil specimens—Darwin went exploring and collecting on several islands. He had developed a good eye for subtle differences among creatures. He noticed, for instance, that the mockingbirds on different islands had slightly different markings. He learned, too, that the giant tortoises on various islands had different-shaped shells (Figure 3.1, top).

Darwin did not yet draw any conclusions about such things. After five weeks of baking on the black rocks of the Galápagos, he and the *Beagle* sailed west. A stop in Sydney provided some semblance of the comforts of home and a chance for Darwin to roam inland. On a stroll at dusk along a chain of ponds in the Blue Mountains, he was lucky to spot several platypuses. A semiaquatic animal with fur like a beaver, a bill like a duck, and one of only two species of mammals that lays eggs, the first specimens sent back to Europe were so perplexing they were taken as fakes. At the very least, the platypus proved that its Creator had a sense of humor (Figure 3.1, bottom).

Darwin was struck by both the contrasts and similarities between Australian animals and those elsewhere. He saw marsupial carnivores that naturalists had dubbed "tigers" and "hyenas" because they were superficially similar to their namesakes in Asia and Africa. And he watched a "Lion-Ant" catch insect prey in its conical sand trap, just as he had seen in Europe. Darwin interpreted the similarities as the work of one Creator.

Voyaging further west to South Africa, Darwin took the opportunity to visit the famous astronomer Sir John

FIGURE 3.1. Galápagos birds and platypus

Top, Darwin found three very similar but distinct mockingbird species on four different islands. From Darwin (1838–1841). Reproduced with permission from John van Wyhe (2002).

Bottom, the platypus, one of two egg-laying mammal species remaining on the planet. From Lydekker (1904).

Herschel in Cape Town. Darwin had earlier devoured Herschel's *A Preliminary Discourse on the Study of Natural Philosophy* (1831) and was thrilled to meet the author. Widely considered the greatest scientific mind of his time, Herschel was also keenly interested in geology, fossils, and botany. He personally cultivated more than 200 plants species on his property and had noted how his South African plants seemed to be graduated with respect to one another, with some species appearing to fill links between others. Unbeknownst to Darwin, in the months before their meeting, Herschel had been thinking and corresponding about what he called "the mystery of mysteries"—the replacement of extinct species by new species.

Darwin recorded the meeting in his diary as "the most memorable event which, for a long period, I have had the good fortune to enjoy," which said a lot coming from someone in the fifth year of an around-the-world voyage. Whatever was said between the astronomer and the young naturalist, Darwin would soon begin to see his specimens in a different light. On the voyage home, while organizing his notes on the birds of the Galápagos, Darwin returned to the slightly different mockingbirds and tortoises that occur on different islands and noted how they have essentially identical habits. What could explain these facts? According to special creation, God made a different species for each island. But another possibility occurred to Darwin: maybe the animals were all variations of one type. He jotted:

"If there is the slightest foundation for these remarks, the zoology of Archipelagoes—will be well worth examining; for such facts would undermine the stability of species."

Species might change. It was Darwin's first eureka moment. A small, tentative step onto the slippery slope of evolutionary thinking. Darwin was not an expert in any field and did not know the identity of most of his specimens. He could not be sure whether certain Galápagos animals were different species, or variations of the same species. He was counting on experts back home to sort through and help make sense of all of his specimens.

And help they did. During the winter following Darwin's return, leading experts pored over his specimens and some collected by other shipmates. The fossils he collected in South America, for example, were identified as giant, extinct mammals with close anatomical relationships to living armadillos, llamas, rodents, and sloths.

Darwin was most stunned to learn that twenty-five of twenty-six Galápagos land birds were not only distinct species but unique to the islands. He would later remark, "I never dreamed that islands about fifty or sixty miles apart, and most of them in sight of each other, formed of precisely the same rocks, placed under a similar climate" would have different species.

Darwin was gripped by the possibility that species did in fact change.

He opened a series of private notebooks and jotted his thoughts down in stream of consciousness fashion. His first surge of ideas was about the genealogy of species. He reconsidered the great differences between Australian mammals and those elsewhere and inferred that those differences could be the result of the long separation of their respective continents. The giant fossils he found buried in South America seemed to be larger, extinct versions of living mammals, which suggested that living species are descended from older extinct types. He started imagining life as being organized like an irregularly branched tree, with some branches dying and others forming anew. He suspected that the similarities he saw between animals in any one region were because they sprang from one branch. And then on page 36 in his notebook, he wrote "I think" and doodled a diagram (Figure 3.2).

It was a simple, crude sketch, but Darwin's drawing was a radical new picture of life, a genealogy of species in which one species gives rise to new and slightly different species, which in turn give rise to grand-species, and then great-grand-species and so forth. Darwin's bold idea was that species are born from existing species just as naturally as children are born from their parents. The revelation marked the end of special creation for Darwin.

He turned next to thinking about how new species form. In the fall of 1838, he happened to read for pleasure an essay from forty years earlier by the economist Thomas Malthus *Essay on the Principle of Population*. Malthus emphasized how populations could outgrow their food supply, resulting in poverty, famine, and death. Darwin was well aware of how the same forces acted in nature and that many plants and animals produced far more offspring than could survive.

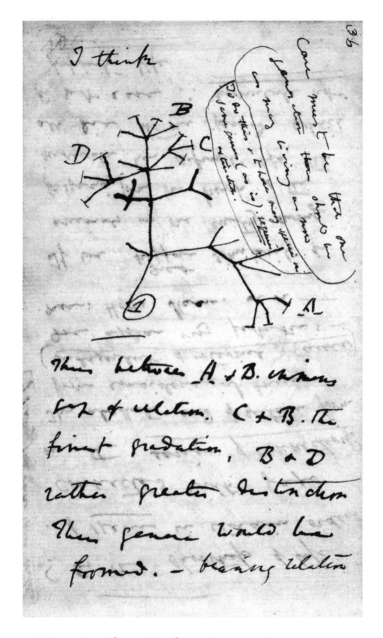

FIGURE 3.2. Darwin's new view of nature

Darwin's first drawing of a tree of life in which species are descended from existing species. From his notebook B. Reproduced by kind permission of the Syndics of Cambridge University Library. DAR.121.

And he knew well from his collections how individuals varied. The thunderbolt struck again. "It at once struck me that under these circumstances favourable variations would tend to be preserved, and unfavorable ones to be destroyed. The result of this would be the formation of a new species," he later recalled. Darwin would dub this preservation and destruction of variations "natural selection."

Just two years after his epic voyage, the twenty-nine-year-old Darwin had a theory to explain the origin of species by natural selection. His theory, however, would not see the light of day for twenty years.

Darwin's reasons for not publishing for so long have been grist for generations of scholars. At the very least, Darwin initially believed that it would be premature. He would need more evidence, much more evidence, to answer his own questions and to overcome the doubts that were certain to come from the scientific establishment and beyond. And so, as he later explained to a friend, "I determined to collect blindly every sort of fact, which could bear any way on what are species."

For fifteen years, Darwin gathered all sorts of facts and observations that bore on his theory. He was in the meantime more prolific than any naturalist in his day, producing a popular travelogue of his voyage, nine volumes on the geology and zoology of the places he visited on the *Beagle,* a new and ultimately correct theory on the formation of coral reefs, and four volumes on barnacles—as well as fathering ten children!

Finally, in 1855, he wrote to his cousin, "I am hard at work at my notes collecting & comparing them in order in some two or three years to write a book with all the facts & arguments, which I can collect, *for & versus* the immutability of species." He added, "Far the greatest fact, about myself is that I have at last quite done with the everlasting Barnacles."

At last, he was ready.

DARWIN'S OTHER BIRDS

Well, almost ready.

Darwin understood his challenge in persuading others of his theory was partly one of imagination. If he was correct, one ancestral species, such as a finch, could give rise over time to a myriad of species. But the making of such diversity required an unknown, perhaps immense amount of time, and the steps in between were not apparent. How could he demonstrate even the plausibility of many different forms descending from one ancestor? And, how could he show that natural selection was powerful enough to shape life's diversity?

His answer: Pigeons!

A pigeon-breeding craze had swept across England. Since the birds were inexpensive to keep and bred easily, the hobby was open to all regardless of class or income. The many breeds called pouters, runts, toys, carriers, fantails, and tumblers differed in the color, number, and pattern of feathers as well as the size and shape of their bones and beaks. They were so different, in fact, that breeders believed that the different types were descended from different wild species. Darwin suspected, however, that they were all descended from one species, the rock pigeon (*Columba livia*) (Figure 3.3).

Darwin thus saw the astonishing range of pigeon diversity as a perfect surrogate for a group like the Galápagos finches, with the hand of breeders acting in an analogous way to nature in selecting the various forms.

Visitors to Darwin's stately home in the village of Down no doubt expected to find the great naturalist in his study poring over notes and books, absorbed in his thoughts. Instead, from May 1855 on, they were perhaps more likely to find him admiring his pigeons in the birdhouse he had built in the back of his garden—or boiling their carcasses or measuring their skeletons.

FIGURE 3.3. Wild rock pigeon and fancy pigeon breeds

Composite of drawings of pigeon breeds from Darwin's *Variation in Animals and Plants under Domestication* (1868); the wild rock pigeon is at upper left. Image by Paul D. Stewart/SCIENCE SOURCE.

The usually reclusive Darwin started attending pigeon shows and even joined two breeding clubs. Just as he had done with fellow naturalists for fifteen years, he milked his new acquaintances for all of the facts they could provide about their beloved breeds. His own flock grew to a peak of almost ninety birds, and his love for the animals grew accordingly. He invited the eminent geologist Charles Lyell to visit, "I will show you my pigeons! Which is the greatest treat, in my opinion, which can be offered to human beings."

Darwin was particularly impressed by the breeders' eyes for subtle features in selecting which birds to mate, tiny details that escaped the untrained eye. He learned that traits that defined each breed—beak length, the number of tail-feathers, etc.—offered plenty of variability. The breeders did not seek large or sudden changes in size or shape; they sought more subtle changes in the overall forms of the birds, with criteria for some traits specified to within one-sixteenth of an inch.

In his own breeding experiments, Darwin took an approach that was taboo among the pure breeders—the cross-breeding of different types. He discovered that all hybrids were completely fertile. And when he crossed white fantails with black barbs, and then crossed the hybrids to one another, he obtained some blue birds with markings very similar to wild rock pigeons! The two results proved to Darwin beyond any doubt that all breeds were descended from one common wild ancestral species.

His decision to devote several years to studying pigeons paid off. He was much better armed to deflect the doubts of naturalists. Since the birth of his theory, Darwin had shared his views with a few trusted colleagues and sounded out the opinions of many other naturalists without divulging his own. Despite the rapidly growing knowledge of natural history, Darwin reported, "I . . . never happened to come across a single one who seemed to doubt the permanence of species."

On the strength of his pigeon work, Darwin would write, "May not those naturalists . . . learn a lesson of caution, when they deride the idea of species in a state of nature being lineal descendants of other species?"

One exception among those naturalists would emerge in 1858, when Alfred Russel Wallace sent Darwin a short manuscript with virtually identical ideas as his species theory. Wallace had spent four years exploring the Amazon, survived the burning and sinking of his ship home, and then journeyed throughout the Malay Archipelago. Wallace had observed very similar patterns of facts as Darwin: slightly different species on nearby islands; much variation among individuals of the same species; the production of far more offspring than could possibly survive; and he had also read Malthus.

Wallace's paper and a short excerpt from Darwin were published together in 1858 in a professional journal to which almost no one paid any attention. It wasn't until the next year when Darwin completed and published his opus with the full title *On the Origin of Species by Means of Natural Selection, or the Preservation of Favoured Races in the Struggle for Life* that the world took notice.

The very first chapter of one of most important books ever written devoted no fewer than ten pages to pigeons (and contrary to most impressions, none in the entire book to finches!).

GOOD HEAVENS: THE POWER OF NATURAL SELECTION

The *Origin of Species* was not the finish line for Darwin or his theory, it was the beginning of a new leg of his journey. Having finally divulged his ideas and evidence, they were opened to scientific and public scrutiny. Naturalists zeroed in on the issues that they found unpersuasive, or unacceptable. Darwin's explanation for the natural (not divine) origin of species drew mostly hostile reviews in both scientific journals and the popular press.

A few scientists, such as Asa Gray, a prominent Harvard botanist and devout Presbyterian, sought to reconcile their theistic views with the new Darwin/Wallace theory. Darwin carried out a long, voluminous, and very warm correspondence with Gray, who was a strong proponent of Darwin's theory and arranged the first American publication of *The Origin of Species*.

"I had no intention to write atheistically," Darwin confided to Gray. "But I own that I cannot see, as plainly as others do, & as I sh^d wish to do, evidence of design & beneficence on all sides of us. There seems to me too much misery in the world. I cannot persuade myself that a beneficent & omnipotent God would have designedly created the Ichneumonidæ [a family of parasitic wasps] with the express intention of their feeding within the living bodies of caterpillars, or that a cat should play with mice."

Darwin viewed natural selection as the "paramount power" and continued to seek more evidence of its action. Gray provided generous help to Darwin's newest fancy—the study of orchids. After completion of *The Origin,* Darwin hurled himself into a comprehensive work on the plants, just as he had done before with barnacles and pigeons. Per his usual approach, he milked botanists around the world for information and specimens. And per his usual genius, he made yet more penetrating discoveries that bolstered his case.

Darwin was fascinated by the beautiful plants and paid particular attention to how the plants were cross-fertilized by pollen. He was struck to discover the various contraptions— "contrivances" he called them, derived in different species from different parts of the plant that were used for enticing insects into their flowers and attaching pollen to them.

Darwin's discovery of the "the endless diversity of structure ... for gaining the very same end" created an opportunity to make what he described as a "flanking movement" on his critics. He would appeal to readers' interest in the beautiful and popular plants and lead them away from special creation and toward an evolutionary explanation. Why would an Omnipotent Creator bother making so many different modifications for exactly the same purpose, Darwin asked, when natural selection on different variations in different plants could do the job?

Just before completing his book *On the Various Contrivances by which British and Foreign Orchids are Fertilised by Insects* Darwin received a specimen called the Madagascar star orchid (*Angræcum sesquipedalia*) that took pollination contrivances to a whole new level. He wrote excitedly to his English botanist friend Joseph Hooker, "I have just received such a Box full from Mr Bateman with the astounding Angræcum sesquipedalia with a nectary a foot long—Good Heavens what insect can suck it."

Darwin discovered that the flowers of the beautiful and rare plant produced a very sweet nectar, but only in the lower inch and a half of a nearly 12-inch long, whip-like, green spur extending below the flower petals, a nectary far longer than any other orchid he'd seen (Figure 3.4). By probing the long spur with a narrow tool, he discovered that the pollen became attached only when he reached the bottom of the nectary. Having studied the ways various insects were recruited to carry pollen, Darwin surmised that during feeding, some moth would only pick up pollen when its tongue reached the distant nectar. "What a proboscis the moth that sucks it, must have!" Darwin told Hooker. No such moth was known, but Darwin ventured a prediction in his book, "In Madagascar there must be moths with proboscis capable of extension to a length of between ten and eleven inches."

Such an extreme character in the star orchid was to Darwin an astonishing example of the accumulative power of natural selection to modify a trait in any direction, and to any length. Just as breeders were able to produce extreme forms—fancy pigeons, greyhounds, longhorn cattle, etc.—through artificial selection, Darwin was confident of natural selection's ability to produce extreme creatures—frilly woodpeckers, long-necked giraffes, and orchids with foot-long nectaries. So, why not moths with preposterously long tongues?

The orchid and its hypothetical moth prompted Darwin to propose a kind of arms race to explain their extreme features: natural selection on the moth would favor a longer proboscis in order to reach the orchid's nectar, while natural selection on the orchid would favor a longer nectary so that the moth would have to fully insert its proboscis and thereby make contact with pollen.

Darwin's confidence in natural selection was borne out. Forty-one years after his prediction, and twenty-one years after his death, a moth with an 11-inch long proboscis, *Xanthopan morganii praedicta*, was discovered—in Madagascar (Figure 3.4).

Darwin's idea of an arms race driving longer tongues and flower tubes turned out to be just as prescient. More than a century later, careful studies of wild populations of a hawkmoth species and iris plants in South Africa have revealed considerable variation in tongue length and flower-tube length, and demonstrated that longer-tubed plants are in fact more successful at spreading their pollen via longer-tongued moths, and thus favored by natural selection.

FIGURE 3.4. Star orchid and moth

Left, the star orchid (*Angraecum sesquipedale*) has an unusually long nectary that contains pollen. Right, the moth *Xanthopan morganii praedicta,* whose existence was predicted by Darwin decades before it was discovered, has an unusually long proboscis with which it reaches the nectar, and thereby picks up pollen. Photo courtesy of Robert Clark.

Moths are not alone among nectar feeders in having extremely large mouthparts. The sword-billed hummingbird has a 4-inch (10 cm) long bill, the mega-nosed fly has a 5.7 cm proboscis, and the nectar bat of the Ecuadoran cloud forest has an 8.5 cm tongue—which is one and a half times longer than its entire body.

Good Heavens, those animals really can suck it.

But perhaps more to Darwin's and our point, if natural selection can produce such ridiculously long bills, noses, necks, and tongues, is a large-brained ape any greater feat?

DARWIN AND CHANCE

Skepticism about natural selection, however, was only half the battle. Darwin asserted that natural selection operated on the variation that existed among individuals of a species. The source of that variation was the focus of considerable commentary and controversy among both foes and allies alike. Darwin plainly admitted that neither he nor anyone else knew the direct cause of variation, or the laws of inheritance. However, at many points throughout *The Origin*, Darwin alluded to the chance or accidental nature of variation.

Darwin clearly grasped that there was an element of probability in the appearance of variations. He pointed out why domestic breeders kept large herds or flocks, "[A]s variations manifestly useful or pleasing to man appear only occasionally, the chance of their appearance will be much increased by a large number of individuals being kept." He reasoned the same would be true in nature: "forms existing in larger numbers will always have a better chance, within any given period, of presenting further favourable variations for natural selection to seize on, than will the rarer forms which exist in lesser numbers."

In several instances, he explicitly described variations as "accidents":

"I can see no reason to doubt that an accidental deviation in the size and form of the body . . . so that an individual . . . would be able to obtain its food more quickly, and so have a better chance of living and leaving descendants."

"[W]ith animals, as with plants, any amount of modification of structure can be effected by the accumulation of numerous, slight, and as we must call them accidental, variations, which are in any manner profitable."

Critics pounced on the implications of accidental variation—the lack of any role for a Creator or intelligence. These were yet more grounds for many to reject Darwin. Accidental variations also did not sit well with Gray, who favored a process guided in some way by a Creator. Designed or created variation did not sit well with Darwin who thought that the abundant variation he described had dispelled any need for divine intervention. Darwin dismantled Gray's reasoning with pigeons and woodpeckers:

"Asa Gray and some others look at each variation, or at least at each beneficial variation . . . as having been providentially designed. Yet when I ask him whether he looks at each variation in the rock-pigeon, by which man has made by accumulation a pouter or fantail pigeon, as providentially designed for man's amusement, he does not know what to answer; and if he, or any one, admits [that] these variations are accidental, as far as purpose is concerned . . . then I can see no reason why he should rank the accumulated variations by which the beautifully adapted woodpecker has been formed, as providentially designed."

An understanding of the cause of variation and the role of chance, however, would be even more elusive than a monstrous Madagascan moth. Not for another century could anyone say what variations were, and only in the last few years could we catch chance red-handed.

CHAPTER 4

RANDUM

"Despite the constant negative press covfefe"

—DONALD TRUMP, TWITTER, MAY 31, 2017

I FIRST STARTED READING baseball box scores when I was eight. Every morning, I would dash out to pick up "The Times" from our front steps—no, not *The New York Times*, in my neck of the woods that would be *The Toledo Times*—to see what every player did in each game. Batting averages were tabulated only in the Sunday paper, so to stay on top of matters I, like most normal kids, calculated the statistics for my favorite players every day throughout the week.

On Friday June 14, 1974, I flipped past all of the coverage of the Watergate scandal to get to the important stuff, only to be disappointed that Thursday had been a day off for most teams. My disappointment, however, did not last long.

The Detroit Tigers—the nearby major league team with a large Toledo fan base—had played an exhibition game for charity against the Cincinnati Reds. The game did not count in the league standings or the stats, but I clipped out the short account of the game and have saved it for nearly half a century. See if you notice anything:

Tigers Edge Reds In Exhibition

DETROIT (AP) — Detroit got three runs on a pair of early homers and added a big seventh ining to defeat Cincinnati of the National League 5-3 in their inter-divisional Sandloot Benefit exhibition baseball game Thursday night.

Detroit opened up the scoring in the first inning when Mickey Stanley walloped a bases empty shit, and added a pair in the second on a two run blast by Jim Northrup off starter Tom Hall.

FIGURE 4.1. Toledo Times, June 14, 1974

There are three mistakes, but it was the one typo of an "i" in pace of an "o" that caught my eye. I thought it was hilarious (still do).

At least I assume it was a mistake. I doubt some anonymous sports writer thought, "No one is going read this so I will pull a fast one." Although I have since checked several other Ohio papers that carried the exact same story and discovered that in their versions Mickey Stanley hits a bases empty "shot."

Anyway, I doubt anyone got fired.

On the other hand, after Queen Victoria visited the now-historic Menai bridge in Wales and *The Times* of London reported: "The Queen herself pissed graciously over the magnificent edifice," I suspect the Palace was not amused. Nor was the White House smiling when the *Washington Post* reported after President Wilson's evening at the theatre with his bride-to-be Edith Galt, "The President gave himself up for the time being to entering his fiancée." The Post had the edition recalled but some copies managed to reach the streets.

One has to wonder after tales of Kennedy, Clinton, and Trump escapades whether this 1631 version of the King James Bible is inside a White House nightstand:

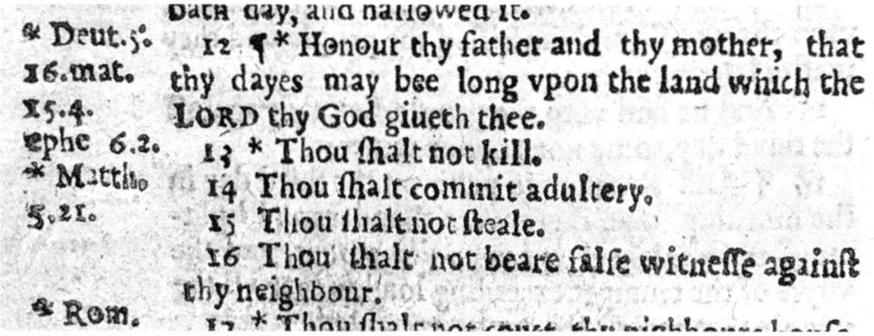

bath day, and hallowed it.

* Deut.5.
16.mat.
15.4.
ephe 6.2.
* Matth.
5.21.

12 ¶ * Honour thy father and thy mother, that thy dayes may bee long vpon the land which the LORD thy God giueth thee.
13 * Thou shalt not kill.
14 Thou shalt commit adultery.
15 Thou shalt not steale.
16 Thou shalt not beare false witnesse against thy neighbour.

* Rom.
17 * Thou shalt not couet thy neighbours house

FIGURE 4.2. The Commandments of the 1631 King James Bible

Courtesy of University of Leicester Library, (SCS03550).

The seventh commandment (line 14 above) was not the only howler. Deuteronomy 5:24 offered this gem: "And ye said, Behold, the LORD our God hath shewed us his glory and his great-asse," Glorious, no doubt, but the preferred spelling was "greatnesse."

The blasphemy was not detected for a year. King Charles I was, well, royally pissed. He ordered that all copies be burned and revoked the printers' license, one of whom died in debtors' prison.

What a difference a single letter or word can make.

The same is true in life's alphabet. While many typos in the genetic text do not change its meaning, or produce a bit of harmless gibberish, some changes can have large effects on how creatures look, operate, or behave. Some for better, some for worse.

Here is an example of a typo in a tiny part of a genetic text that recently made history:

The original text was this: KKKYMMKHL
One typo changed it to . . . :   KKKYRMKHL

That mistake, a change of M→R, has killed more than 35 million people.

How can such a small change be so deadly? I'll get to that a bit later.

The crux of the matter is the cause of that change. Monod argued fifty years ago that all changes in the genetic text, called mutations, are a matter of chance—*accidents that occur at random, regardless of their potential consequences.* All of his philosophical conclusions hinge on the veracity of that assertion.

I say assertion because what made Monod's claim particularly bold was that there was very little *direct* knowledge of DNA mutations at the time. In 1970, scientists had no access to the genetic text of living things, no way of determining the sequence of DNA. As you have no doubt heard, the situation has changed dramatically so that today every thing's and everybody's DNA is an open book, even people or creatures that are dead or extinct.

That hard-won ability gives us the power to test Monod's assertion, to trace typos in the library of life, to spot mistakes as they are made, while even more powerful technology enables us to peer deep into the machinery of life and, finally, catch chance in the act.

ONE IN A BILLION

To test the case for the randomness of mutations, we should first have a good idea of conditions that need to be met. There are three levels to be considered—randomness at the level of a population of organisms, randomness at the level of DNA, and randomness at the level of mechanism:

In a population, a particular mutation would be said to arise at random if: i) it arises without respect to its consequences, and ii) the individual(s) in which it occurs cannot be known or predicted beforehand.

Within individuals, if mutations are random, they would be expected to occur in a random distribution across DNA.

Within DNA, a random mutation would be generated by a mechanism governed by chance.

Monod had evidence at the first level (and only the first level).

Soon after the development of the first powerful antibiotics such as penicillin, clinicians and researchers encountered the phenomenon of antibiotic-resistant bacteria, where some patients treated with these new wonder drugs developed infections that were no longer susceptible to them. The isolated resistant bacteria were found to be stable generation after generation, demonstrating that resistance was due to some mutation that was inherited.

Since these antibiotics were new chemicals to which the bacteria were unlikely to have ever been previously exposed, the origin of the resistance mutation was a puzzle. Did the presence of the drug somehow direct the formation of mutations that made the bacteria resistant? Or, did resistance mutations occur at random—regardless of whether the drug was around, and those lucky bacteria were able to grow in its presence.

It was difficult to distinguish between a directed or random explanation until the brilliant team of young bacteriologists Esther and Joshua Lederberg came up with a clever and elegant experiment. The Lederbergs realized that if they could devise a way to test bacteria that had never been exposed to an antibiotic for resistance, they would have the answer to whether the antibiotic was necessary for the formation of mutations. The challenge was that resistance mutations were very, very rare, perhaps just one in one billion or more bacteria. How could they find, let alone study, a needle in such a large haystack?

The properties of bacteria are often inspected by growing them on a solid medium such as agar (a gelatin-like substance made from seaweed) infused with nutrients. The bacteria form colonies that are scattered across the surface of the agar plate. The Lederbergs' key invention was a way of making exact replicas of plates containing millions of bacterial colonies by pressing the plate against a patch of sterilized velveteen fabric, and then pressing the fabric against the surface of a fresh plate. The fabric picked up enough bacteria that multiple replicas could be made of the original plate.

To trace the origin of antibiotic resistance mutations, the Lederbergs first grew the original bacteria in the absence of the antibiotic, then plated replicas on antibiotic-containing plates. They found that a few antibiotic-resistant colonies grew on the antibiotic-containing plates. These mutations could still have been directed somehow by the antibiotic. However, because of the replica technique, the Lederbergs could go back to the original plates and prove that the resistance mutations existed in the original colony *even though those bacteria had never been exposed to the antibiotic* (Figure 4.3).

Using the same technique the Lederbergs also showed that other traits, such as bacterial resistance to viral infection, also arose by chance and did not depend on prior exposure to the virus or any specific environment.

The Lederbergs' now-classic experiments were about as low-tech as one could imagine. They required some petri dishes, some cloth, and some simple nutrients. Probably fifty bucks worth of materials in today's currency.

But here is the biggest kicker to the story: The work was accomplished in 1951, before the structure of DNA was solved, and when there were still some lingering doubts

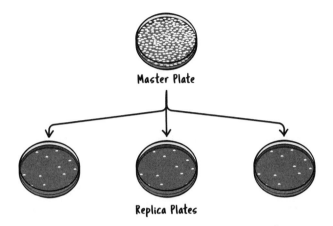

Master Plate

Replica Plates

FIGURE 4.3. Demonstration that mutations occur at random in a population

The Lederbergs grew a master plate of bacteria in the absence of antibiotic, then replicated that population of bacteria on plates containing antibiotic. The few antibiotic-resistant colonies that grew occurred in the same place on different plates, demonstrating that the original colonies on the master plate had evolved resistance in the absence of antibiotic, that is, at random. Illustration by Kate Baldwin.

about the role of DNA in inheritance. No one had any clue at the time what a mutation actually was in physical reality.

That was about to change.

EUREKA!

The breakthrough was one of the extremely rare moments in science when answers to several fundamental questions arrived in one fell swoop. I am going to walk through a couple of key details because they will resurface shortly in the crucial issue of randomness.

Francis Crick and James Watson pursued the mystery of the structure of DNA off and on, mostly off, from 1951 to 1953. After several failures, and with some crucial clues obtained from Rosalind Franklin's X-ray pictures of DNA crystals, Watson hit upon the key to the structure one Saturday morning in February 1953 when he was playing with cardboard models of the structures of the four chemical bases—adenine (A), guanine (G), cytosine (C), and thymine (T)—that make up DNA. The long stumbling block was figuring where and how the bases fit in the overall structure. Watson had tried schemes of pairing "like with like"—A with A, G with G, and so forth—but the resulting molecule was an ugly, impossible mess.

Working alone that morning, Watson tried something new by arranging the bases in different pairs: the large base A with the small base T, and the large base G with the small base C. When these specific pairs were placed adjacent to one another, Watson saw that chemical groups on one base were able to form a type of bond called hydrogen bonds with chemical groups on the other base (Figure 4.4, left).

Bingo. Watson instantly realized that the bonds between the A-T and C-G base pairs could hold together the two chains of the double helix. And, because the pairs were the same size and shape, they could neatly stack on top of each other in the interior of the long double helix, like steps in a winding staircase (Figure 4.4, right).

Watson and Crick quickly recognized that the base-pairing rules (A with T, C with G) solved three puzzles all at once. First, the job of DNA in inheritance was to faithfully pass genetic information from generation to generation. The base-pairing rules meant that the sequence of bases on one chain was the complement of bases on the other chain, and therefore determined the sequence on the other chain. Thus, a sequence of bases could be faithfully copied generation after generation according to the rules.

Second, for DNA to be the genetic text in all living things, and yet for each species to be different from one another, the molecules of DNA must differ somehow. Watson and Crick realized that because the base pairs fit so neatly in the interior of the long double helix structure, countless per-

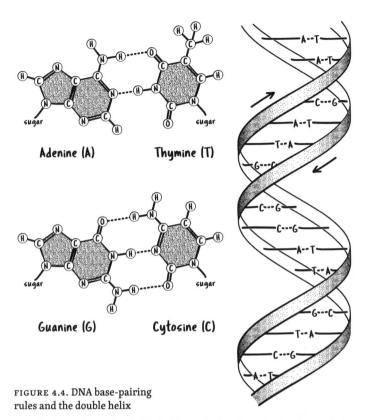

FIGURE 4.4. DNA base-pairing
rules and the double helix

Left, the bases A and T, and C and G form similar shaped pairs that are held
together by hydrogen bonds. Right, the double helix model of DNA in which
two strands of DNA are held together by hydrogen bonds between bases on
opposing strands. Illustration by Kate Baldwin based on data from Watson
(1980).

mutations of base sequences (ACGTGATCGATTACA ... etc.)
must be possible. The precise sequence of bases would carry
the genetic information specific to each species—the library
of the entire biosphere written in a four-letter alphabet.

And finally, for changes to occur in inherited traits,
changes must occur in the DNA text. Watson and Crick proposed that mutations were due to a change in the sequence
of bases.

Breathtaking. Three deep insights into the workings of life from the chemistry of what scientists had once thought was just a boring, do-nothing polymer.

Watson took a celebratory trip to Paris and shared the story with Monod, who was astounded.

The great leap forward spurred many new questions. The first was how the four-letter DNA alphabet works to specify the characteristics of living things. Over the next decade, a small cadre of researchers cracked the genetic code by determining how the sequence of bases in DNA encodes the instructions for making proteins, the molecules that do all the work in every creature and determine most traits (more about that in the next chapter).

It was on the basis of this new biochemical knowledge and experiments such as those by the Lederbergs that Monod asserted that all innovation and change in the biosphere was due to random changes in the genetic text, a matter of chance. It was a reasonable inference, but one based entirely on indirect evidence.

The second major spate of questions spawned by the double helix are at the heart of the direct, physical evidence for randomness: How is the text copied? How often do mistakes occur? And, the ultimate question, *why* do mistakes occur? Biologists have made spectacular strides in answering these questions with remarkable precision and detail.

THE FASTEST TYPISTS

The task of copying the DNA text is the job of a family of enzymes called DNA polymerases (because they make polymers of DNA). These molecular machines are the fastest and most accurate typists on Earth.

Consider the job of copying the genetic text of just the simple bacterium Escherichia coli (E. coli) that lives in our guts. E. coli has a single circular chromosome that is about 4.6 million base pairs around and contains more than 4,000 genes. The bacterium can double as quickly as every thirty minutes, which means 4.6 million bases have to be copied in that time, or about 150,000 bases per minute. Amazingly, E. coli starts the process at just one place in its DNA and works in both directions (clockwise and counterclockwise) at the speed of about 1,000 bases per second. To put that in perspective, the fastest human typist Stella Pajunas was clocked at 216 words or about 1,000 characters per minute on an IBM electric typewriter in 1946. That is the world record; the average professional typist works at about one-third this speed and could copy this book in about 16 hours. The E. coli DNA polymerase could crank it out in about five or six minutes.

Now let's talk about accuracy. A pro is expected to be about 97 percent accurate, meaning about three typos in every 100 characters. Detailed studies of DNA polymerases reveal that the enzymes make as few as one mistake in 10,000 to 100,000 bases, for 99.999 percent accuracy.

That impressive number only tells part of the story. Those rates are measured using isolated enzymes in a test tube. The actual mutation rate inside bacterial cells is much, much lower—just one or two mistakes in 10 billion bases per generation. This is because the copying machinery also includes a proofreading function that corrects misinserted bases; that process increases accuracy by another 100-fold over just the copying function. And there is another biochemical machine that searches for mismatches in the newly copied DNA that improves accuracy another thousand-fold. Multiplied together—1 mistake in 100,000 made by the enzyme, 1 in 100 of those slipping through proofreading, and 1 in 1,000 of those escaping mismatch detection—and we get just one mistake in $100,000 \times 100 \times 1,000 = 10$ billion bases.

That's all well and good for bacteria, you might say, but what about the mutation rate in humans? We have a very accurate measure of mutation rates in humans because we can now determine the sequence of anyone's entire DNA (their *genome*) cheaply and quickly. By sequencing the DNA of parents and their children, we can count the exact number of mutations that occurred in one generation by spotting the bases that are different in the children from either of their parents. The range is about 40–70 new mutations within every child's roughly 6 billion base pairs of DNA, for a rate of about one mutation per every 100 million base pairs. That's higher than the rate in bacteria but still pretty low.

I'll say more about those mutations in Chapter 6, but before you panic about yourself or your kids (or future kids), it is important to mention most of these new mutations will be neither good nor bad and will have no effect at all. This is because there is a lot of open space in our genome between genes, and most mutations will occur in those spaces. And even if one mutation lands in a gene and disrupts it, most of our genes exist in two copies, and the one good copy will get us by.

The 40–70 new mutations within every child are contributed by both Mom and Dad, via the egg and sperm, respectively (Don't make me explain how the egg and sperm get together!). Amazingly, biologists have pushed the sequencing technology so far as to be able to identify mutations in single individual sperm cells. Stephen Quake, a Stanford University biologist who pioneered the exploration of what he dubbed "the ejaculome" discovered 25–36 mutations per sperm cell, consistent with expectations of roughly half the total number of mutations that are found in children.

With such powerful technology for finding mutations within individuals' DNA, we can now ask a fundamental question about randomness. Are the mutations randomly distributed across the genome? Such studies have been conducted in a variety of species and the answer is, to a first approximation: "Yes!"

So far, so good for Monod. We now know how often mistakes arise and that they are randomly distributed. But one question still lingers, why do mistakes happen at all?

That answer has come only very recently, by capturing chance in the act.

QUANTUM FIBRILLATION

Back in 1953, when Watson and Crick were working on the structure of DNA, they were initially tripped up by the then-limited understanding of the chemistry of the four bases. Watson copied the chemical structures from a textbook but some were incorrect (a good lesson for all of us!). Fortunately, another scientist working in their group straightened him out, and history was made.

It turns out that those chemical details are not only important to solving the riddle of DNA, they are pivotal to the question of random mutation. So, hang in there, here comes a little chemistry. Feel free to skim through what follows; you can get the gist of it without taking a chemistry class. Or you can dig in and use it to impress your friends (or your pastor!).

Many molecules, including the four bases that occur in DNA, exist in alternative forms called tautomers that can shift from one form to the other. The difference between these forms involves the relocation of a hydrogen atom (a proton) on the ring structures of the bases. That shift affects which groups are available for hydrogen bonding with other bases. One form of the bases guanine and thymine is called the "keto" form; the alternative is the "enol" form (Figure 4.5). What first tripped Watson up was that he copied down the enol forms. His colleague pointed out that the keto forms are the far more common forms, and that correction led to Watson's eureka.

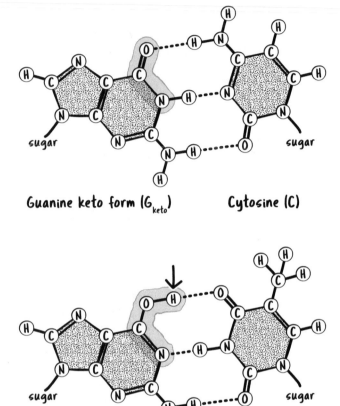

Guanine keto form (G_{keto})   Cytosine (C)

Guanine enol form (G_{enol})   Thymine (T)

FIGURE 4.5. A fleeting shape-shift at the root of mutation

Comparison of the typical G-C base pair involving the keto form of guanine (G) (top) with the more transient enol form that can make a G-T base pair (bottom) and cause a mutation. The difference is due to the transient shift in the position of a hydrogen atom on the larger guanine ring (arrow). Illustration by Kate Baldwin based on data from Bax et al. (2017).

There was an important insight, however, in Watson's initial mistake. He and Crick realized that the enol forms could form hydrogen bonds, *but with the wrong base*—a G with T or an A with C. In their initial reports, they suggested, "spontaneous mutation may be due to a base occasionally occurring in one of its less likely tautomeric forms" by a hydrogen atom changing its location. They imagined that if a base such as G was in the rare form when the DNA was being copied, the wrong complementary base could get inserted into the double helix such as a T where a C should be.

Fast-forward 60 years and that is exactly what biochemists have been able to capture. The coup was made possible by very sophisticated technologies for observing molecules at the atomic level. The events turn out to be extremely difficult to detect because of the fleeting moment of the shift to the rare form—lasting less than one one-thousandth of a second before the molecule shifts right back to the common form. But biochemists have managed to capture the moment and in essence take snapshots of transient mispairs and of DNA polymerases caught in the act of incorporating the wrong base. This fleeting shape-shift within DNA bases accounts for more than 99 percent of all misincorporation mistakes.

These discoveries reveal that the event at the root of mutation, the source of all diversity in the biosphere, is an inescapable, fundamental matter of physics—a quantum transition between chemical states—a chance fibrillation at the atomic level.

Mutation, then, is a feature, not a bug in DNA.

In every organism, in every cell, whenever DNA is copied, changes will occur because of the intrinsic characteristics of the very bases that endow DNA with its properties. Mutation, change, is unavoidable and inevitable.

Now let's see what beauty, what complexity, and a little later, what troubles chance can create.

CHAPTER 5

BEAUTIFUL MISTAKES

"Name the greatest of all inventors. Accident."

— MARK TWAIN

THE WORLD'S REACTIONS to the facts of evolution have spanned the extremes. One impulse is to banish them, as codified for example in State of Tennessee House Bill No. 185:

> That it shall be unlawful for any teacher in any of the Universities, Normals and all other public schools of the State which are supported in whole or in part by the public school funds of the State, to teach any theory that denies the story of the Divine Creation of man as taught in the Bible, and to teach instead that man has descended from a lower order of animals.

Passed in March 1925, the statute was the basis for the famous trial of high school teacher John Scopes, who was convicted. The law was upheld as constitutional in state court and not repealed until 1967.

If that seems outrageous, wait until you hear about the other end of the spectrum. The very same year of the Scopes trial (1925), top officials of the Soviet Communist Party agreed to finance an expedition to western Africa by zoology professor Ilia Ivanov. The scientist had made his reputation as a pioneer in the artificial insemination of livestock, where sperm is obtained from selected donor males and delivered mechanically to females. The process had not been widely adopted until Ivanov's methods, which were used successfully on thousands of horses and sheep. By circumventing the mating process, Ivanov also believed that artificial insemination would allow the creation of hybrid animals, and he was able to produce some novel varieties such as zorses (horse-zebra) and zubrons (bison-cow).

Ivanov's new project was also aimed at creating a novel hybrid, only this time between humans and chimpanzees—a "humanzee." Ivanov believed that the two species were closely related enough that they might make a viable hybrid, which he thought would provide powerful confirmation of the evolution of humans from apes.

Official Soviet support for the project was largely ideological. The humbly-titled Commissariat of Enlightenment described the research as an "exclusively important problem for Materialism" (the doctrine that only matter exists) while others hoped it would undermine religious teachings and the power of the Church. If successful, the hybrids would also be a great publicity coup for Soviet science.

Ivanov had more than Soviet support. The leadership of the Pasteur Institute in Paris, then perhaps the most eminent biomedical research institution in the world, encouraged the research and offered Ivanov access to chimps at their facility in French Guinea. The endeavor and its potential significance were also widely publicized. The *New York Times* reported that the experiment aimed "to support the doctrine of evolution, by establishing close kinship between man and the higher apes." The press coverage aroused much interest, but also some detractors. During a stay in Paris, Ivanov received a threatening letter from the white supremacist Ku Klux Klan (KKK) which he took as positive evidence of his work's "exceptional scientific, but also . . . social significance."

The logistics of the experiment turned out to be more of a problem than the KKK. Ivanov's first trip to French Guinea in 1926 was a bust. He discovered that all of the chimpanzees at the facility were too young and sexually immature. Ivanov traveled again to Guinea in 1927 and helped with the capture of several adult chimps. At last, Ivanov managed to inseminate three female chimps with the sperm of a human donor. However, they did not become pregnant.

Ivanov returned to the Soviet Union with more than a dozen chimps to continue his efforts but this time in the opposite direction—the insemination of human females with ape sperm. After two years of official deliberation, Ivanov secured permission to seek "no less than five" volunteers with the right idealistic (not financial) motives, who would agree to live in isolation at a Soviet primate research station for at least a year under a doctor's care.

Ivanov got his volunteers, but by then all of his imported apes had died in transport or captivity and the experiment had to be postponed. In the meantime, other events overtook Ivanov as a purge of Soviet ranks unfolded. Along with many others, the scientist was arrested by the secret police on trumped-up charges and exiled to Kazakhstan where he died in 1932. The Ivanov episode was subsequently buried in Soviet memory.

Fortunately for the remaining chimps in the world, the proof of evolution has not required their mating with humans. We are long past the question of whether humans or other species evolved through natural means. The legitimate and burning questions today are about *how* species evolve:

How do new capabilities and lifestyles arise? And, how do new species form?

These questions are at the heart of the origin of creativity and innovation in the biosphere. You may be surprised to learn that some of the greatest minds in biology have wrestled with the relative contributions of natural selection and mutation to invention, and that a lively debate continues to this day. Darwin vested great and primary creative power to natural selection: "I can see no limit to the amount of change, to the beauty and infinite complexity of the coadaptations between all organic beings, one with another and with their physical conditions of life, which may be effected in the long course of time by nature's power of selection." But Monod, with full knowledge of a century of Darwinian thinking, asserted that random mutation, and therefore "chance alone is at the source of every innovation, of all creation in the biosphere."

So, let's dare to ask, which of these two bona fide geniuses is right? Or more right?

The answer matters because the greater the creative role of random mutation, the more life appears, per Monod, to be driven by chance.

Fortunately, there is an important new source of insights that springs from our recent ability to peer into any species' DNA and pinpoint the precise cause of some new capability or adaptation. We'll get better acquainted with DNA by first peeking into one of Darwin's favorite creatures.

PIGEON FEATHERS

The English trumpeter, the Indian fantail, the Old German owl, the Old Dutch capuchin, and the Jacobin are different breeds of pigeons that appear to be strikingly different to the untrained eye (Figure 5.1). But aficionados recognize one trait they have in common—a whorl or crest of feathers

FIGURE 5.1. Crested pigeons

From the left, the Jacobin, Old Dutch capuchin, and English trumpeter all bear feathers that form a crest around their necks and heads. Photos courtesy of Michael D. Shapiro.

on their heads. Most breeds—the English pouter, the English tumbler, the racing homer, etc. as well as the ancestral wild rock pigeon, lack such crests.

So, a simple question is: How did certain pigeons get their fancy crests?

For the first 150 years since Darwin kept his birds, finding a precise answer was not at all simple. Breeding experiments revealed that the presence or lack of a crest was determined by a single gene. However, identifying that one specific gene among the thousands in any bird's DNA, and figuring out how that gene is different in crested and uncrested birds, is a proverbial needle-in-a-haystack problem. Such information was largely out of reach until only very recently, when much faster and cheaper DNA sequencing methods put the DNA sequences of every creature within our grasp.

In order to find the important needles in a haystack of pigeon or any species' DNA, we have to have a grasp of the language of DNA, and of how DNA information is decoded in making the working parts of living organisms. You can learn the language of DNA; it has a very small alphabet, a very limited vocabulary, and its rules of grammar are simple. The payoff for learning this language is being able to understand not only the source of innovation and biodiversity, but the source of human individuality and the causes of disease that I tackle in the next two chapters. You might want to bookmark this short section for later reference.

Let's start with chromosomes, genes, and DNA. Every organism's hereditary information is carried by one or more *chromosomes* inside its cells. Every chromosome contains a long molecule of DNA; some of yours are more than 200 million bases long. Each *gene* occupies its own interval along a DNA molecule (Figure 5.2). Individual chromosomes may contain thousands of genes.

Recall from Chapter 4 that DNA is made of four distinct *bases* that are abbreviated by the single letters A, C, G, and T. The most amazing fact about DNA is that all of life's diversity is generated through the permutation of just these four components. The two strands of DNA are held together by strong bonds between pairs of bases that lie on opposite strands; A always pairs with T, and C always pairs with G. It is the unique order of the thousand or more bases in a segment of DNA (ACGTTCGATAA, etc.) that determines the unique information encoded within each gene.

Using just these four bases, the entire DNA of each species encodes thousands of different *proteins*, which are the molecules that do all of the work in our cells and bodies—from carrying oxygen, to breaking down food, to copying DNA for the next generation. Proteins themselves are made up of building blocks called *amino acids*, of which there are twenty different types (for shorthand, these are also represented as single letters in a protein sequence). The chemical properties of these amino acids, when assembled into chains averaging about 400 in length, determines the unique activity of each protein.

The relationship between the DNA code and the unique sequence of each protein is well understood because biologists cracked the genetic code 50 years ago. The decoding

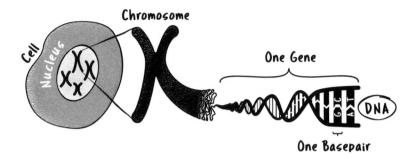

FIGURE 5.2. The relationships between chromosomes, DNA, and genes

Chromosomes are found in the nucleus of cells; each chromosome contains one long molecule of DNA, and each long molecule of DNA contains many genes that span many base pairs of DNA. Illustration by Kate Baldwin.

of DNA in the making of proteins occurs in two steps. DNA is first *transcribed* into a single strand of messenger RNA; this RNA is the complement of one strand of the DNA. The messenger RNA is then *translated* into proteins. The genetic code is read (from the RNA transcript) three bases at a time, with one amino acid determined by each triplet of three bases (Figure 5.3).

There are 64 different triplet combinations of A, C, G, and T in DNA, but just twenty amino acids. Multiple triplets code for particular amino acids (and three triplets code for nothing and mark the stopping point in the translation of proteins—just as periods mark the end of a sentence). Much to our convenience, and a testament to life's common origin, this code is the same in every species (with a few minor exceptions). Given a specific DNA sequence, it is straightforward to decipher the protein sequence that DNA sequence encodes. Similarly, given various kinds of mutations in DNA—substitutions, insertions, and deletions—we can predict exactly how protein sequences are altered (Figure 5.4).

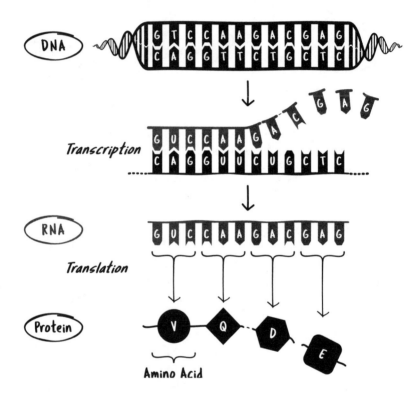

FIGURE 5.3. How information in DNA is decoded

The decoding of DNA into protein occurs in two steps: DNA is first transcribed into a single strand of messenger RNA that is a complement of one DNA strand; then the messenger RNA is translated into a sequence of amino acids that form a protein. Illustration by Kate Baldwin.

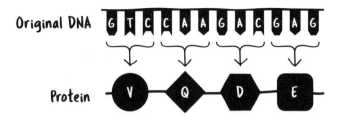

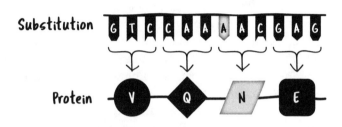

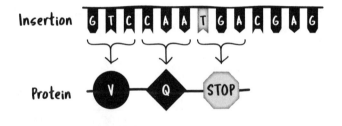

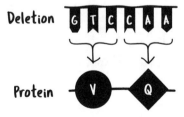

FIGURE 5.4. Different kinds of mutations modify the sequence of DNA

Substitutions, insertions, and deletions commonly occur in DNA that alter the resulting protein sequence. Illustration by Kate Baldwin.

The needle in the haystack challenge comes from the size of genomes. The pigeon genome contains 2.6 *billion* base pairs (our genome has about 6 billion) and over 17,300 genes. Nevertheless, biologist Mike Shapiro at the University of Utah and a team of collaborators from across the world have been able to figure out exactly how pigeons got their crests.

To do so, they examined and compared the DNA of a broad sample of 22 crested and 57 uncrested breeds. They discovered that the presence versus absence of a crest is determined by one specific gene, and most interestingly, by just one difference at one specific position in the sequence of that gene. The uncrested birds have a C:G base pair where the crested birds have a T:A base pair. This means that a mutation occurred in an ancestor of the crested forms that changed a C to a T. The gene encodes a protein called the ephrin B2 receptor; the one base difference in DNA in turn causes one difference in the protein's sequence in which the amino acid arginine is replaced by cysteine (Figure 5.5).

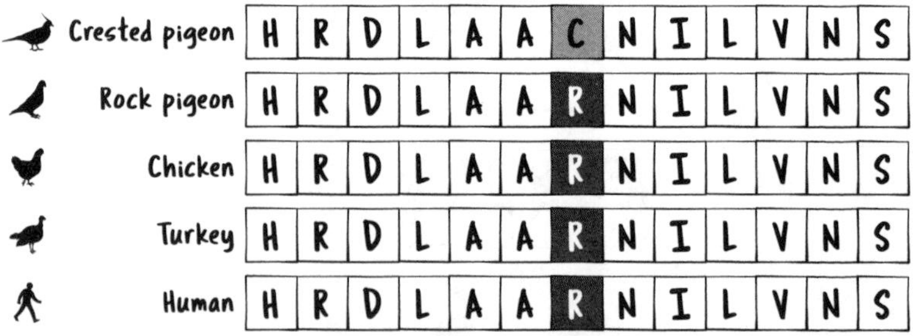

Crested pigeon	H	R	D	L	A	A	C	N	I	L	V	N	S
Rock pigeon	H	R	D	L	A	A	R	N	I	L	V	N	S
Chicken	H	R	D	L	A	A	R	N	I	L	V	N	S
Turkey	H	R	D	L	A	A	R	N	I	L	V	N	S
Human	H	R	D	L	A	A	R	N	I	L	V	N	S

FIGURE 5.5. A single mutation creates the pigeon head crest

A portion of the Ephrin B2 protein sequence from several species is shown using one-letter abbreviations for amino acids. Crested pigeons have a mutation that causes a single change from arginine (R) to cysteine (C) at one position. Illustration by Kate Baldwin based on data from Shapiro et al. (2013).

Shapiro and his collaborators also discovered how that single change makes such a big difference in the bird's appearance. The mutation causes the polarity of feathers on the head and neck to be reversed such that they grow toward the top of the head instead of down the neck, and so form a crest around the head.

Now let's consider how the origin of the pigeon's crest bears on the issue of the contributions of mutation and natural selection to invention. The origin of the crest is exquisitely clear: a mutation occurred at some time in an ancestral pigeon that generated a crest. The presence of the crest in a wide variety of breeds is explained by breeders *selecting* for the crested trait in the development and propagation of these breeds.

So, which is the inventor—mutation or selection?

Darwin believed that natural selection acted gradually on "infinitesimally small" variations. He and later generations of biologists assumed that sufficient variation was always present in any given population to modify a trait in any direction. In this view natural selection initiates evolutionary change and is creative, whereas mutations merely provide "raw material" upon which natural selection acts.

But the crest was not present in the ancestral rock pigeon. Rather, the reversal of the pigeon head feather pattern and the formation of a crest was brought about by a single specific mutation in one gene, in one step, not many small steps. There was nothing "infinitesimally small" or "raw" about the effect of the ephrin B2 mutation.

This sort of discovery—of specific mutations responsible for discrete changes in creatures—now fills the pages of scientific journals, and warrants a rethinking of the potential creative role of mutation. As we did in the last chapter for the randomness of mutation, we need some criteria for evaluating the creativity of mutation. A dictionary definition of invention is the creation of something that did not exist before. I suggest that a mutation should be deemed creative if any of the following criteria apply:

1) The mutation creates a new physical trait
2) The mutation creates a new molecular function or capability
3) The mutation creates a new physiological function of a gene

The feather crest mutation meets the first criterion. Interestingly, Shapiro and his team also searched for the gene and mutation responsible for the formation of a head crest in domesticated ringneck doves, a species that diverged from pigeons over 20 million years ago. Lo and behold, they found that the responsible mutation is in the very same ephrin B2 receptor gene as the pigeon crest mutation, but at a different base within the gene. The discovery suggests that the invention of a crest in these birds can occur in only very limited ways.

But two feather crests do not make a general rule. Let's look at some more inventions and ask what mutation can or cannot do. I have told tales of our chaotic physical planet—of rising mountains, freezing oceans, and the fibrillations of the Ice Ages. So, we'll peer into the DNA of some of the fascinating creatures that inhabit (or inhabited) some extreme environments and see how they managed to do so.

SMALL STEPS FOR DNA, A GIANT LEAP FOR WOOLLY MAMMOTHS

One evening in 2001, biologist Kevin Campbell was watching a television documentary about the unearthing of a woolly mammoth from the permafrost of Siberia. A simple question also occurred to him: How did these Ice Age animals cope with the severe cold?

The superb fossil record of mammoths reveals that their ancestors arose in equatorial Africa about 7 million years ago and that they colonized high latitudes during the early stages of the Ice Age, about 1–2 million years ago. Well-preserved, mummified specimens display several anatomical adaptations to the cold. Unlike their Asiatic and African cousins which have to dissipate heat, mammoths had several features that helped to conserve heat in the frigid north—thick fur, skin glands that kept their woolly coat well-oiled, much smaller ears, and shorter tails.

But Campbell was especially curious about what he could not see—the physiological mechanisms that enabled the animals to stand on frozen tundra. He knew that Arctic animals such as reindeer and musk ox reduce heat loss through their thinly insulated extremities by keeping them cold, just barely above freezing. They avoid frostbite because their blood vessels are arranged in a counterparallel network, such that arteries running down a leg transfer heat to veins running up the leg. The lower temperatures in the extremities, however, make it more difficult for hemoglobin, the protein that carries oxygen from the lungs through the blood to body tissues, to deliver its vital cargo.

Campbell wondered whether there was anything special about mammoth hemoglobin. There was just one small obstacle to finding out—the species was extinct and had been for about 10,000 years. There was no mammoth blood to study.

But these were the early days of ancient DNA technology, when researchers began to retrieve and analyze tiny amounts of fragmented DNA from mummies and fossils. Campbell teamed up with Australian biologist Alan Cooper and Michael Hofreiter, an ancient DNA expert then at the Max Planck Institute in Leipzig, Germany, to see what could be obtained from a 43,000-year-old mammoth thigh bone discovered in the permafrost of northern Siberia.

Using a technique called polymerase chain reaction (PCR) the researchers were able to make copies of the mammoth's versions of the two genes that encode the two different chains of the hemoglobin protein. They spotted three differences in one chain that had arisen since mammoths split off from their Asian and African relatives. Hemoglobin is one of the best-studied proteins in biology. Campbell and his colleagues recognized that at least two of the differences involved positions in the chain that rarely varied among mammals.

This observation was just a correlation. To test whether mammoth hemoglobin functioned differently, the researchers had to bring the protein back from the dead. So they engineered bacterial cells to make the mammoth hemoglobin and compared its properties with that of Asian elephant hemoglobin. They found out that mammoth hemoglobin was indeed much better at releasing oxygen at lower temperatures than elephant hemoglobin.

There are other circumstances in addition to extreme cold where hemoglobin's job is especially challenging. At high altitude, the percentage of oxygen in inspired air is the same but the partial pressure of oxygen which drives gas exchange in the lungs is much lower than at sea level. For example, it drops by about 50 percent at an altitude of 5,500 m (17,000 feet), which creates the risk of oxygen deficiency (hypoxia) for animals and humans.

Many species of birds, however, are able to fly (a very energy-consuming process) at very high altitudes. Perhaps the most famous is the bar-headed goose which migrates from India over the Himalaya en route to Mongolia and has been documented at over 21,000 feet. Similarly, the Andean goose lives at altitudes up to 18,000 feet. Examinations of the hemoglobin genes and proteins of these and a variety of other high-altitude birds have revealed a number of specific mutations that increase the protein's affinity for oxygen relative to the hemoglobins of low-altitude birds, and thereby help to saturate arterial blood and provide sufficient oxygen to body tissues.

In these birds and the woolly mammoth, one or a few mutations were sufficient to impart new properties to hemoglobin that helped the animals expand their ranges into very challenging environments. Score these mutations as creative. These simple substitutions in existing genes are just one type of mutation; however, there are other types of mutations with much more dramatic effects on DNA and creatures.

A MATTER OF DEGREES

The adaptations of woolly mammoth and bar-headed goose hemoglobins are nifty, but sometimes, extreme circumstances require even more extreme measures.

Take Antarctica, for example. As I discussed in Chapter 2, temperatures across the planet have dropped a great deal over the past 50 million years. Antarctica was once ice-free and lush green, and surrounded by waters that while not tropical, were at least moderate. Today, however, if you happen to go on an Antarctic cruise and are tempted to take a plunge—don't. The sea temperature around the continent is about minus 1.9°C (28.6°F). You would quickly become a human ice cube.

But there is life in those icy waters, lots of life. It was a big surprise to some of the first explorers that the coldest waters in the world are in fact teeming with fish. And that poses a scientific mystery: Fish from tropical or temperate waters freeze at about minus 0.8°C, so something must enable Antarctic fish to cope with colder waters.

That something is antifreeze.

Examination of Antarctic fish blood serum revealed that it does not freeze until it reaches about minus 2.1°C, the reason being that it is chock-full of proteins that work as antifreeze.

The main enemy of fish in these waters is not so much the cold, but ice. The water contains small crystals of ice that, should they enter the fish through its gills or be swallowed, would nucleate the formation of larger ice crystals and— bam! They're fish sticks.

What the antifreeze does is bind to ice molecules and prevent them from growing into larger, fatal crystals. The blood serum of temperate fish lacks these proteins, so that tells us that antifreeze is a new invention. In fact, different groups of polar fish have different antifreeze proteins, so it is clear that fish have invented antifreeze more than once. Antifreezes present great opportunities, then, for exploring how new things are invented.

Eelpouts are one such group. They are not handsome fish, indeed some might say butt-ugly, but you have to admire their ability to thrive 500–700 meters below the surface of the icy waters of McMurdo Sound. Researchers Christina Cheng and Art DeVries and collaborators from the Chinese Academy of Sciences asked a very simple question: How did the eelpout antifreeze arise? There are two possibilities: the antifreeze could be something entirely new, or it could be something old doing a new job. The answer turns out to be a bit of both and reveals how mutations can create new genes and functions in a few large steps.

A key clue came when it was noticed that the antifreeze protein bore a remarkable resemblance to a section of an- other protein found in other fish, and even mice and hu- mans, which means that this other protein has been around for a long time. The protein is an enzyme that is involved in the making of a specific type of sugar called sialic acid that is often attached to molecules on the surface of our cells. The enzyme, called sialic acid synthase (SAS for short), is much larger (about 360 amino acids) than the antifreeze (about 65 amino acids), but the antifreeze sequence is very similar to a 65 amino acid sequence at the tail end of SAS (Figure 5.6, top).

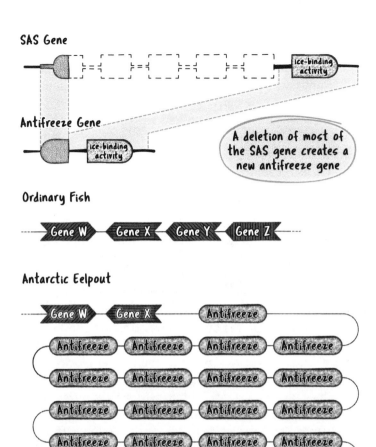

FIGURE 5.6. The invention of antifreeze

Top, A deletion within the SAS gene (dotted boxes) in an Antarctic fish left a segment encoding a protein with ice-binding activity fused to the front of the original gene, creating an antifreeze gene. Bottom, This antifreeze gene was subsequently duplicated many times such that the Antarctic eelpout has about thirty copies of the gene nested among other genes (W, X, Y, Z), while ordinary fish lack antifreeze altogether. Illustration by Kate Baldwin, based on data from Deng et al. (2010).

The reason for this strong resemblance was deciphered by some expert sleuthing through the DNA of eelpouts and other fish. That detective work revealed that the antifreeze gene evolved from a chunk of the SAS gene (Figure 5.6, top). That chunk encoded a bit of protein that, on its own, had some ability to bind to ice crystals. That activity came in handy as the temperature of the ocean fell. Since the birth of the antifreeze protein, it has undergone many changes that have increased its ice-binding power.

In addition, the antifreeze gene has been greatly expanded in number, which enables the fish to make *a lot* of antifreeze. The eelpout has over thirty copies of the gene, all tandemly arrayed next to one another on a chromosome (Figure 5.6, bottom). The tandem arrangement of those genes told the researchers that the eelpout's battery of antifreeze genes was built up over time by another well-known mutational mechanism in which whole genes spanning thousands of base pairs, or even larger blocks of DNA containing multiple genes, are duplicated in a single step.

One more important point to the story. The difference between the ancestral sugar-synthesizing gene and the newer antifreeze genes revealed that one critical step in the origin of the antifreeze genes was the deletion of a piece of the ancestral gene. In this case, a *deletion* was a key creative mutation.

The making of antifreeze, then, involved several large steps. First, there was a duplication of the SAS gene. Next, a deletion occurred of most of the new SAS gene, leaving the chunk. The chunk was then duplicated, and its duplicate duplicated, and so forth again and again over the course of millions of years of fish evolution (see Figure 5.6). These mutations generated a new kind of molecule (an ice-binding protein) with a new physiological function (preventing freezing), which meet the criteria (2) and (3) mentioned previously. This phenomenon of novel and expanded sets of genes doing new jobs is widespread across all forms of life. Other novelties such as venoms that enable snakes to capture prey and the milk that mammals feed to their young have similar evolutionary stories in that they are derived from genes that had other functions.

One could write a whole book on the creativity of mutations. I'll spare you and let these few examples suffice to illustrate ways in which various kinds of mutations can be creative.

I say "can be" because most mutations are not creative. In fact, most mutations have no impact for several reasons. First, there is a lot of DNA sequence in animal and plant genomes that has no essential function (perhaps as much as 95 percent of our DNA). Mutations that occur in these tracts are generally of no consequence. Second, because of the redundancy of the genetic code, mutations within genes that change a DNA base do not necessarily change the sequence of the encoded protein. About three-quarters of all mutations that are found within the coding parts of regions do not alter the "meaning" of a triplet because the original and mutant triplet are synonyms that encode the same amino acid. And third, even mutations that do cause changes in protein sequences may not have any functional effects, or may have a detrimental effect.

Creative mutations, then, are a small minority, and rare. As I explained in the previous chapter, any mutation is rare at any specific position in DNA, arising once in about 100 million individuals (depending on the species). Gene duplications and deletions are similarly rare. Nonetheless, like an amateur golfer who, given enough swings will eventually hit the target, given enough pigeons, doves, fish, etc. over generations, a particular mutation will occur in a population.

But there is a rub.

The mammoth's hemoglobin was not, mutationally speaking, one lucky shot but involved multiple mutations. Because mutations are independent events, the probability of two mutations occurring at once in the same gene is vanishingly small. It is the Kim Jong-Il multiple holes-in-one problem—the probability of two mutations is the multiple of the probability of each independent mutation, about 1 in 10^8 × 1 in 10^8 = 1 in 10^{16} or 10 *quadrillion* individuals. To put that number in perspective for woolly mammoths, that would require more than a billion times the entire biomass of all animal life on Earth.

Impossible?

Nope. Cue natural selection.

CLIMBING STEPS

In the woolly mammoth, researchers found that at least two of the three changes in the mammoth hemoglobin independently affect oxygen delivery. It is reasonable to infer that each of these changes provided some incremental benefit to mammoths in improving their performance in the cold and expanding their range. But given the improbability of two mutations happening at once, how could the two mutations (or three) wind up in *all* woolly mammoths?

The process is stepwise, not instantaneous.

If one mutation provides a significant performance advantage, then over time it can increase in frequency in a population by increasing the reproductive output and survival of those that inherit it versus those that don't. This is the competitive process of natural selection. For example, if a mutation confers just a 3 percent advantage, meaning that those with it produce 103 viable offspring for every 100 produced by those lacking it, the mutation can become ubiquitous in a large population in fewer than 1,000 generations.

If and when a second beneficial mutation occurs, then it will arise in an animal that already possesses the first mutation and can subsequently spread as well. This is the process of *cumulative* natural selection. It works like climbing a set of stairs, with new mutations providing each new step up

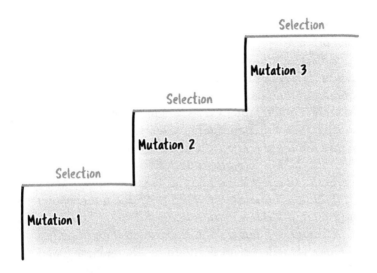

FIGURE 5.7. The staircase of evolution

Multiple changes can occur in a gene through a cumulative process. New mutations provide each stepwise change in a gene (the rise in the stair) while natural selection spreads each mutation in a population (the run of each stair). In this manner, multiple changes can accumulate in a gene. Illustration by Kate Baldwin.

(the rise in the stair), and natural selection providing each step forward by spreading the mutation in a population (the run of the stair) before the next step up occurs (Figure 5.7). The process takes a long time in slowly-reproducing creatures, but we can see this happen in real time in rapidly multiplying creatures like microbes and viruses, and cells in our own bodies. Take a good look at Figure 5.7; you are going to see this set of stairs again.

This staircase draws a picture of the respective jobs of mutation and natural selection, and of what each can and cannot do. Natural selection can't invent anything on its own; invention—a step up—requires an enabling mutation. On the other hand, because a new mutation occurs in an individual, mutation alone cannot change a population nor produce multiple changes at once.

So, chance invents, and natural selection propagates the invention.

Or not. It is very important to appreciate that a given mutation may or may not confer an advantage in a given population at a particular time at some place on Earth. For example, one of the exact same mutations that enabled woolly mammoth hemoglobin to deliver oxygen at low temperatures has also been found in a few individual humans. But in these people, it causes a mild anemia and so it does not spread in the population. Similarly, a mutation that makes fur white on a mammal may be beneficial in snowy regions and a liability elsewhere.

What is good then for one creature is not necessarily good for another; it depends upon their respective circumstances. And what determines those circumstances? As we saw in Chapters 1 and 2, these are largely the external, physical conditions that are in turn shaped by a myriad of contingencies—by chance.

So chance invents, and the fate of that invention depends upon circumstances shaped by chance.

We are a long, long way from Providence, and I don't mean the capital of Rhode Island. But there is still more. The impact of chance in the living world extends beyond the invention of traits to the phenomenon most dear to Darwin—the origin of species. To explore that realm, and close this chapter, let's return to Ivanov's quest and the matter of chimps and humans.

AN ACCIDENTAL TREE

What would have happened had Ivanov succeeded in impregnating a woman with chimpanzee sperm? Would a viable humanzee result?

We can't say for sure, but there are several kinds of facts to consider in weighing the possibility.

Species are defined as reproductively isolated populations. There are two categories of barriers to the formation of hybrids between species: pre-mating factors, such as animal behavior or anatomy, that preclude mating; and post-mating factors largely genetic in nature that prevent viable (or fertile) hybrids from developing. It is well established that genetic "incompatibilities" can build up over time between two species that may preclude the development of viable hybrid offspring. These include not only the kinds of mutations we have encountered thus far that affect the sequences of genes, but larger-scale rearrangements in chromosomes—inversions, translocations, breakages, and fusions—that may not affect gene sequences but can change the order and location of genes on chromosomes as well as chromosome number.

Ivanov bypassed the anatomical and behavioral obstacles to chimp-human mating by artificial insemination, so the question boils down to genetics: Can the human genes contributed by the mother work together with the chimp genes contributed by the father and produce an infant?

From sequencing each species' DNA, we know that across most of our genome, chimpanzee DNA sequences are about 98.8 percent identical to that of humans. That means that in the approximately 6 million years since our respective lineages split, about 12 single-base mutations have occurred on average in either the chimp or human lineage every 1,000 base pairs; about 35 million mutations in total across 3 billion base pairs. It is also the case that the chimp's genes are spread across 24 pairs of chromosomes, while we have 23 pairs.

Those figures indicate both a great deal of similarity and a lot of differences that could tilt the viability of a humanzee in either direction. The question is what extent of genetic differences is tolerable in an animal hybrid, or conversely, what extent is a barrier to hybrid formation?

Evolutionary biologists S. Blair Hedges and Sudhir Kumar and their colleagues at Temple University in Philadelphia have scrutinized the issue across the tree of life. They have uncovered a striking consistency in the time required for complete speciation among animals such as mammals and birds, of about 2 million years. Beyond this time appears to be a "point of no return" after which reproduction between any two lineages is barred by the buildup of genetic incompatibilities.

In short, the great, branching tree of life appears to be the inevitable product, or by-product, of the steady accumulation of random mutations in populations. It is a profound insight that Darwin could not and would not have imagined. It is worth quoting the authors:

"The lineage splitting seen in trees probably reflects, in most instances, random environmental events leading to isolation of populations, and potentially many in a short time. However, the relatively long [time-to-speciation (2 million years)], a process resulting from random genetic events, will limit the number of isolates that eventually become species. Under this model, diversification is the product of those two random processes, abiotic and genetic."

I'll say it another way: Look around you at all the beauty, complexity, and variety of life. We live in a world of mistakes, governed by chance.

As for the humanzee question, since humans and chimps have been separated for much longer than 2 million years (about 6 million years), my bet is no, a humanzee infant is not likely to be viable.

Even if Ivanov failed to obtain a hybrid, there is one other possible outcome from his experiment that he did not contemplate but that is important to raise, especially in light of what the world is experiencing in 2020: Ivanov could have triggered a deadly pandemic.

In the previous chapter, I mentioned a typo within the sequence of a protein KKKY**M**MKHL that changed it to KK-KY**R**MKHL and that had killed more than 35 million people.

I will reveal here that the first short sequence is part of a protein made by the simian immunodeficiency virus (SIV) that infects chimps, as well as gorillas and various other Old World monkeys. The one change (M→R) in the second sequence is found in all three major strains of human immunodeficiency virus (HIV-1), including the strain responsible for the AIDS pandemic. The mutation enabled SIV to jump from chimps into humans three separate times and become HIV-1.

It is not clear exactly how the first infections occurred. Most speculation focuses on human contact with infected chimp blood or body fluids in the course of hunting, preparing, or consuming bushmeat. The approximate time and place of the first infections, however, are clear and appear to have occurred in west central Africa around the beginning of the twentieth century. HIV-2, the lesser-known evil twin of HIV-1, also originated in west central Africa. The widespread distribution of SIV means that no matter where Ivanov obtained chimp sperm, he was unknowingly running the risk of introducing the virus into humans, and potentially enabling the origin of HIV/AIDS.

The repeated, independent mutation of SIV into viruses that can infect humans, and the spread of those viruses between people, show how the origins of viral pandemics are also a matter of chance. Among randomly mutating viruses in one animal species some may be able by chance to infect humans. If that species just happens to make very close contact with humans, then the virus may infect one or more people, and subsequently spread from person to person.

These "spillover" events of the animal to human transmission of a novel virus were also the genesis of the 1918 influenza pandemic (from birds), the origin of the 2002–2004 severe acute respiratory syndrome (SARS) epidemic (from palm civet), and of multiple outbreaks of Middle East Respiratory Syndrome (MERS) since 2012 (from camels).

Indeed, many of humanity's worst scourges are now understood to be of animal origin, including smallpox (originating more than 2,000 years ago from related viruses carried by rodents) and measles (originating about 1,000 years ago from the rinderpest virus of domesticated cattle).

The SARS-CoV-2 coronavirus that emerged in China in late 2019 and swept the globe has a similar origin. Initial studies of its genome reveal features very similar to those of bat coronaviruses as well as specific mutations found so far only in pangolin coronaviruses (also known as scaly anteaters). The latter are popular items in wildlife meat markets in China and elsewhere. It appears then that SARS-CoV-2 may have spilled over from pangolins into humans.

There is a larger lesson here beyond don't mate with chimps, don't kiss camels, and don't eat palm civets or pangolins: Spillover events, and potential pandemics, are accidents just waiting to happen.

PART III

23 AND YOU

WE NOW KNOW that just like every other species, we humans are here by accident. We emerged through the same interplay of chance-driven external and internal processes that created rock pigeons, woolly mammoths, and eelpouts.

But that "we" is a matter of speaking collectively. When a biologist says "humans," just whom are they talking about? Aren't we each different from one another in various ways? And if so, where does all the diversity within our species come from? What makes each of us unique?

In the next two—and final—chapters, I will make chance personal.

You have probably been avoiding it all of your life, but it's time to think about your parents' gonads, and the moment you were conceived. And to add to your trauma, I'll also mention your grandparents' gonads.

You will see why the K-Pg asteroid and our species' emergence from the chaos of the Ice Age are just half of the story of your good fortune. A whole lot of chance events had to happen in certain ways to put you, and I mean *you* specifically, here on Earth.

And it's not just a matter of chance that you were born, it's chance that helps to keep you alive, and it's chance that might very well kill you, although if you read on, you will see there are ways to reduce that chance.

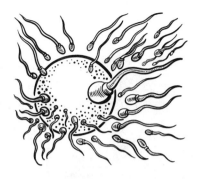

CHAPTER 6

THE ACCIDENT OF ALL MOTHERS

"So remember, when you're feeling very small and insecure,
How amazingly unlikely is your birth;
And pray that there's intelligent life somewhere out in space,
'Cause there's bugger all down here on Earth!"

— MONTY PYTHON, *GALAXY SONG*

FROM THE OUTSIDE, the simple, white, single-story rectangular building looks like many small Christian churches across the rural American South. Inside, it's a typical Saturday night, the service has drawn about twenty parishioners who are joyfully singing a rousing rendition of the gospel hit "Oh Happy Day." But when Jamie Coots, the 42-year-old pastor of the Full Gospel Tabernacle in Jesus Name in Middlesboro, Kentucky, reaches into a small box near the lectern, the celebration takes on a whole new spirit.

FIGURE 6.1. Jamie Coots handling rattlesnake

Reproduced by permission Tennessean\USA TODAY Network.

He pulls out two timber rattlesnakes with his bare hands, gently raises them above his head, and while continuing to sing, calmly walks about the church before passing the writhing serpents to a member of his congregation (Figure 6.1).

The ritual of snake-handling practiced in one hundred or so Pentecostal churches in West Virginia, Kentucky, Tennessee, and other parts of Appalachia stems from a passage in the Bible, Mark 16:18: "They shall take up serpents; and if they drink any deadly thing, it shall not hurt them." While one could argue that the Bible did not specify venomous snakes, that is not how Coots, the third generation in his family to lead the church, interprets it: "To me that's what God taught me to be about."

For Coots and his flock, the handling of venomous rattlesnakes and copperheads is an act of faith—that they won't be bitten, or that if they are bitten, God alone will protect them, as they typically refuse medical treatment.

Coots practices what he preaches. By the time of a national television interview in 2013, he had been bitten nine times. Luckily, some strikes were "dry" bites where the animal does not expel venom. The worst of Coots' injuries was when a timber rattlesnake struck him on his right middle finger. "It was the most pain I have ever felt in my life," he told the reporter. Left untreated, the finger eventually died and broke off.

Why take such risks or endure such pain? "It is an inner peace . . . God has honored me to let me feel his spirit," Coots quietly explained, adding, "If the Bible tells me to jump out of an airplane, I would."

Only three months later, on February 15, 2014, a timber rattlesnake saved him the airfare.

Very few people die from snakebites in the United States. Coots' was one of just three fatalities in all of 2014, but the second of three Pentecostal snake-handlers to die between 2012 and 2015. The main reason there are few fatalities is that most serious bites can usually be successfully treated with antivenom; Coots and the other snake-handlers refused medical treatment. Moreover, most bites are preventable. Reviews of hospital records reveal that the majority of snakebites involve people harassing or handling the snake. As my friend and snake expert Danny Brower was fond of saying, "There are two kinds of snake enthusiasts: Those that have never been bitten, and those that have been bitten *a lot*."

In fact, fatal accidents with any kind of wild animal are extremely rare in the United States. The year 2018 stood out among recent years in having the largest number of people combined (10 total) die from attacks by bears (5), mountain lions (2), alligators (2), and sharks (1). There were only two such fatalities in 2017. Now contrast this with deaths from falling off ladders or scaffolds (569), drowning in swimming pools (723), accidental shootings (486), or poisonings other than by drugs (3,484). The animal encounters make headlines, but as for many chance events, we don't have an accurate sense of the relative likelihood of various accidents. We tend to fear the unlikely and ignore or underestimate the greater threat.

Our poor grasp of probabilities extends to our own individual existence. Our odds of dying from an animal attack are extremely slim, but due to games of chance in our parents' gonads, the odds of our even being alive are truly astronomical. Some of us may have been told that we were unplanned "accidents," but the truth is we are all accidents. What is more, our daily survival depends upon a remarkable system in our bodies that operates on chance.

ONE IN SEVENTY TRILLION

Picture a huge swarm of meteors, perhaps 100 million of them, in the vicinity of a large planet in the solar system. Most are on trajectories that will hurl them into empty space, but a small cluster emerges from the pack and veers toward the giant sphere. Over time, a few close on the target and then appear to bounce off its outer atmosphere. But one penetrates the protective layer and delivers its powerful payload. The large body trembles on impact and emits a shower of chemicals.

But this time, life does not end. It begins.

Perplexed? Try picturing the meteors with heads and long tails.

They are human sperm, and the impact is the moment of fertilization, when a single sperm penetrates the outer covering of a human egg thirty times its size. The trembling and shower are part of the sequence of dramatic physical and chemical changes that occur in the egg that prevent fertilization by other sperm and begin the process of embryonic development.

Out of a swarm of 100 million or more contenders, only a single sperm will swim all the way up the Fallopian tube and successfully fertilize the egg. The fertilized egg is the union of two genomes, half of its chromosomes from the sperm and half from the unfertilized egg. And here is an astounding fact: no two fertilized human eggs will *ever* be the same.

That's right, all that touchy-feely bullshit about how everyone is special actually has some truth to it. Everyone (except for identical twins) is genetically unique—just like snowflakes!

You might be thinking: "Is that really true? I have my mother's eyes, my father's nose, etc. and so does my sister, or brother, etc."

Alright then, time for a pop quiz: If each parent contributes twenty-three chromosomes at fertilization, how many genetically unique children could your parents produce?

Take a guess. 23? 46? 92?

Nope. Here is a hint: The answer is more than 70 . . . *trillion*. That is a 7 with 13 zeros after it.

Let's unpack the math. The key to the calculation is that we are asking how many different *combinations* of chromosomes are possible.

Humans have twenty-three pairs of chromosomes, forty-six chromosomes in all. Twenty-two chromosomes have the same overall structure in boys and girls and are called autosomes, while the other two chromosomes, the X and Y, are the sex chromosomes (boys have one X and one Y; girls have two X chromosomes). Normal mature sperm and egg cells have one copy of each chromosome, so the fertilized egg has the full complement of twenty-three pairs of chromosomes.

Each human then inherits half of their chromosomes from Mom (through her egg) and half from Dad (via his sperm). Mom and Dad, in turn, inherited half of their chromosomes from their Mom and Dad (your grandparents), respectively, and so on. That means that for any given chromosome, there are two alternative chromosomes that could be contributed to a sperm or egg cell. These chromosomes are not identical because they have different ancestries. They will be carrying alternative DNA sequences of most genes.

Sperm and eggs are made through processes that begin with cells that contain pairs of each chromosome and end up with a single copy of each chromosome. Crucially, the process that doles out one of each pair of chromosomes into a sperm or egg is random. That means that every sperm could have one of two versions of each chromosome. The number of different possible combinations of two chromosomes is 4, the number of possible combinations of three chromosomes is 8 and so forth. The number of possible combinations of twenty-three chromosomes is 2 to the 23rd power (2^{23}), or 8,388,608 different sperm. The number of different possible eggs is the same (2^{23}): 8,388,608. The number of possible combinations of sperm and egg is the product of each of these numbers: 8,388,608 × 8,388,608, which equals 70,368,744,177,664 (roughly 70 trillion) different babies.

That is a big number. Testicles try their darnedest, producing about 100 million sperm per day such that a man can produce 2 trillion or more sperm in his lifetime. In contrast, women are born with all the eggs they will ever make—around 1–2 million, which is reduced to about 50,000 by puberty. And by ovulating just one egg per month, they keep family size well below the trillions. After my arrival at 10 lb., 5 oz. (4.67 kg), my mom stopped at four.

The math shows that we are each unique, but it is actually an underestimate of the potential number of unique children. That is because of two more factors that scramble the genetic omelet. The first is a process I have not mentioned

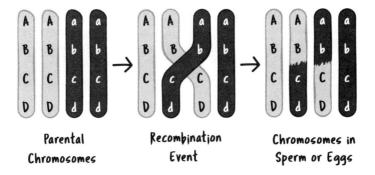

Parental Chromosomes · Recombination Event · Chromosomes in Sperm or Eggs

FIGURE 6.2. Genetic recombination increases the genetic diversity of sperm and eggs

When chromosomes pair during the making of sperm or eggs, they can recombine with each other such that each chromosome is a mosaic composed of genetic information from both chromosomes. Illustration by Kate Baldwin.

thus far called genetic recombination. When chromosomes are paired during the process of making eggs and sperm, they can physically recombine with one another and swap sections. The resulting chromosomes are then a mosaic of the original parental chromosomes (Figure 6.2). Because recombination can happen virtually anywhere along the chromosome, and occurs on average about once per chromosome pair, the actual number of chromosomally distinct eggs and sperm is much larger than 8,388,608. In addition, it is during the making of eggs and sperm when new mutations occur that are not present in either parent. As mentioned earlier, there are about 20–35 new mutations per sperm or egg, and these occur at random throughout the genome, so the number of genetically distinct sperm and eggs is astronomically large.

Thanks to four random mechanisms—the random doling out of chromosomes to sperm and eggs, the random swapping of chromosome bits, random new mutations, and which lucky sperm fertilizes which egg, we are each carrying unique combinations of chromosomes, genes, and mutations. We are each unique accidents—a collision of one genetically unique sperm with one genetically unique egg.

THE LUCK OF THE DRAW

Now that we are all feeling special, time for a few harder facts.

Because all of the processes that produce sperm, eggs, and babies entail large degrees of chance, sometimes those unique genetic combinations are unlucky. About five percent of human babies will have some genetically-determined disorder. Of these about 20 percent are due to a new mutation not present in either parent.

Among the most common genetic disorders are those that are caused by mutations on the X chromosome. These syndromes appear most often in males because they have only one X chromosome, and thus if their only copy of a gene on the X is defective, they will exhibit the syndrome. Females have two X chromosomes so one functional copy of a gene on the other X is usually sufficient to compensate for a defective gene. Hemophilia A, Duchenne muscular dystrophy, and red-green color blindness are all X chromosome-based syndromes that occur predominantly in males and can arise spontaneously from a new mutation not present in either parent. They can also emerge when mothers carrying one defective gene contribute that X chromosome to their sons.

While mutations are random, there are factors that influence the number or kind of mutations that might be present in sperm or eggs. For example, as men age, their sperm carry more mutations on average. This is because the cells that give rise to sperm have undergone more rounds of DNA replication and accumulated more mutations. A recent study of Icelanders and their children revealed that fathers pass on about three additional mutations with every two additional years of age. With more mutations comes a greater risk of a genetic disorder. For example, at least thirty percent of children with autism spectrum disorder (ASD) appear to be due to new mutations. A study of over 5.7 million children in five countries revealed that the risk of fathering an ASD child is about 66 percent higher in men over fifty than men under thirty. Something to think about if you are a supermodel thinking about hooking up with some elderly billionaire.

Another source of genetic disorders comes from the sorting out of chromosomes during the making of mature eggs. Sometimes, both chromosomes of a pair get packaged into an egg along with the usual twenty-two other single chromosomes. If this egg becomes fertilized, the embryo will then have three copies of that chromosome, a condition called trisomy. The most common trisomy is of chromosome 21, which is responsible for Down syndrome, the characteristics of which result from an extra dose of certain genes on chromosome 21.

The risk of Down and other trisomies also rises with age. Eggs are made during a girl's embryonic development, but do not mature until decades later. As women's eggs age, the risk of trisomy becomes about two and a half times greater at age forty than at age twenty-five. Largely because of trisomy and other chromosome abnormalities, women under twenty-four lose about 10 percent of pregnancies in the first trimester, while women over forty-two lose more than half. Trisomies other than chromosome 21 are comparatively rare or never seen in children, which suggests that most babies with other trisomies are not able to make it to term.

There is so much chance involved in our births, we should all feel lucky.

Some more than others.

ONE OF A KIND

Stephen Crohn grew up in Dumont, New Jersey, just across the Hudson River from Manhattan. A baby boomer born in 1947 who came of age in the 1960s, Crohn struggled with his identity as a gay man. Keenly aware of prejudice, he left college to join the civil rights movement and marched with Martin Luther King from Selma to Montgomery, Alabama. An extrovert seeking acceptance, Crohn moved to the Hell's Kitchen neighborhood of New York and joined its still-underground gay culture. He eventually returned to school and developed a career as an artist while working as a copyeditor to make ends meet.

Many years later in 1979, he fell in love with Jerry Green, a handsome and athletic executive chef. They soon moved across country to West Hollywood and indulged in LA's nightlife until Green developed a relentless fever in January 1981. Later that summer, the Centers for Disease Control (CDC) reported unusual outbreaks of pneumonia and a rare cancer in homosexual men in New York and California. By the next winter, Green was blind in one eye and covered with cancerous lesions known as Kaposi's sarcoma. Wasting away before Crohn's eyes, Green died in March 1982, one of the first victims of a disease that the CDC named acquired immunodeficiency syndrome (AIDS) later that year.

The cause of the disease was identified by French and American researchers in 1984. It was a previously unknown virus of a type able to cause cancer in animals, a so-called retrovirus. Most viral infections such as colds or flus are of limited duration because our immune systems mount a vigorous response against them and eliminate them from the body. But the AIDS virus (later named HIV for human immunodeficiency virus) has two particularly sinister properties. First, it infects cells of the immune system, eventually depleting a key cell type necessary for response to many kinds of microbes—leading to pneumonia, fungal infections, and worse. And second, it has a very high mutation rate such that it is continuously changing its appearance to the immune system and evading attack. Doctors and medicines were largely powerless against the disease once it had manifested.

Green was just the first of Crohn's social circle to succumb. Two more died in 1985, another in 1986, and one more in 1989. Crohn feared he would be the next as he had taken no more precautions than any of his friends. Every time he felt unwell, he watched for telltale signs. But as yet still more friends died, those symptoms never came. He was absolutely certain that he had been exposed. Gradually, it dawned on him that he might somehow be resistant to the virus. He started telling doctors and friends that he should be studied.

His hunch was not just a wild fantasy. Medical sleuthing ran in his family. His great uncle was Burrill Crohn, the physician for whom the gastrointestinal disease was named. At family gatherings, he told researchers his story and they agreed that he should be examined.

It was not until 1994 that someone finally did. Dr. William Paxton of the Aaron Diamond AIDS Research Center was looking for men who had been exposed to the virus but not progressed to AIDS, and Crohn's doctor put them in touch. Paxton drew Crohn's blood and exposed a population of his white blood cells, called CD4 T cells, to a massive dose of HIV.

Paxton was astonished, "I couldn't infect his CD4 cells, and that was something which we'd never seen."

The reason why Crohn's T cells were impenetrable took a couple of years to figure out. HIV gains access to cells through two coreceptor proteins called CXCR4 and CCR5 that act like a portal on T cell surfaces. Detailed examination of Crohn's T cells revealed that they did not produce the CCR5 receptor. With part of the portal missing, the virus could not enter his T cells. Sequencing of Crohn's DNA revealed a 32 base pair deletion in *both* copies of his CCR5 gene (denoted as CCR5delta32). This discovery meant that he had inherited the lucky mutation on each of two chromosomes, one from his mother and one from his father.

The discovery galvanized both Crohn and the medical community. Crohn helped to raise money for AIDS research by telling his story in the media and to donors. The medical community jumped on the insight Crohn's cells provided to pursue new kinds of HIV medicines. A decade later, a new drug called maraviroc that blocked HIV's access to the CCR5 receptor was approved for use in patients. That same year, an HIV patient named Timothy Ray Brown received a bone marrow transplant from a donor who carried the same CCR5delta32 mutation as Crohn and became the first person ever cured of AIDS.

Extensive global surveys for the delta32 mutation reveal that it occurs at a significant frequency in Europeans and Eurasians (3–16 percent) but is absent or nearly so from native Africans, South Americans, or Asians. In the populations where the mutation is found, individuals like Crohn who carry two copies necessary for resistance to HIV are relatively rare, typically less than 1 percent of a population.

One of the still-lingering mysteries is why the delta32 mutation exists, because it clearly long predates the origin of AIDS. The delta32 mutation has been found in DNA samples from bodies in central Germany that date as far back as the Bronze Age (2,900 years ago) and was prevalent in the Middle Ages. The HIV virus did not jump from chimps into humans until the early twentieth century (see Chapter 5). One speculation is that the delta32 mutation provided some resistance to some other as yet unidentified pathogen first encountered by people several millennia ago in Europe or Eurasia such that natural selection caused its prevalence to increase. Resistance to HIV in twentieth century people would then just be a fluke.

Whatever its origin, thanks to both of his parents, Crohn had a winning ticket he never expected for a lottery he never

wanted to enter. But the fact is that all of the rest of us fortunate enough to avoid HIV are also kept alive by a part of
our bodies that relies on chance.

THE STAIRCASE OF SELF-DEFENSE

The deaths of millions from AIDS starkly demonstrate the
vital role of the immune system in protecting us from potential pathogens. AIDS patients are ravaged by so-called
opportunistic infections of viruses, fungi, parasites, and
bacteria that are usually held in check in healthy, uninfected
people. The same sorts of infections plague children who
are born with mutations that cripple the development of
the immune system (like the "Boy in the Bubble"), or patients on certain types of chemotherapy that suppress the
immune system.

Every day, we are surrounded, covered, and infiltrated
by potential enemies. By the time we become adults, there
are more bacterial cells in and on our bodies than our own
cells. They are members of about one thousand or so species
that, along with species from about eighty genera of fungi,
make up the human "microbiome." We also come into contact with all sorts of microbes in our environment, in soils
and the foods grown in them, on domestic livestock, and
one another. The awesome power of the immune system
is its ability to respond to virtually any foreign invader—
bacteria, viruses, fungi, or parasites, and to recognize any
foreign protein or carbohydrate by producing antibodies
specific to that foreign agent (called an "antigen"). And, yes,
that includes novel pathogens like SARS-CoV-2.

The potential range of antigens a person might encoun-
ter is so enormous, the immune system would require
millions—perhaps tens or hundreds of millions—of different antibodies to be able to recognize them all. And that
is the crux of one of the greatest mysteries in biology: How

can the immune system recognize and defend against virtually anything that comes its way?

Monod thought that the question was so important that he raised the topic in *Chance and Necessity*. The fundamental mechanism had eluded researchers for decades. Monod did not live to see the solution; the critical breakthrough occurred just after he passed away.

He would have been delighted by the discovery: the immune system is powered by random chance and a staircase very much like the one we've seen before (see Figure 5.7). The rises on the staircase involve certain kinds of mutations that take place only within immune cells; the runs involve selection on these cells. I'll outline the steps on the staircase first, then explain more about how each rise and run works. The big revelation will come again from the math of combinations.

One major branch of the immune response involves a class of white blood cells called B cells (the other major branch involves white cells called T cells). Think of these cells as the body's soldiers in the fight against invaders. B cells undergo a series of steps that eventually leads to their secretion of antibody proteins. Think of these antibody proteins as chemical weapons. They work by binding to a specific antigen; the binding then helps to block, kill, or clear the foreign agent from the body.

B cells are found primarily in the spleen and lymph nodes where they occur in massive numbers. An early step in the B cell immune response is the activation of a B cell when it encounters an antigen that is recognized (bound) by an antibody receptor on the B cell surface. Only a very tiny fraction of individual B cells will recognize any one antigen. When they do, they are triggered to proliferate and greatly expand their numbers (one cell can give rise to 4,000 cells in about a week). This process is called "clonal selection and expansion" (the terms will make more sense shortly) and forms the first run in the staircase (Figure 6.3).

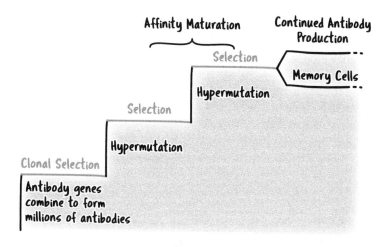

FIGURE 6.3. The staircase of self-defense

Rounds of random genetic events and selection generate antibodies to specific antigens. The first rise is the assembly of antibody genes by combining gene segments at random in developing B cells. The first run is when individual B cells bind to antigen; these clones are then activated and multiply in a process called clonal selection. Within these clones, antibody genes undergo further rounds of mutation (hypermutation; the second and third rises); those of higher affinity for the antigen are selected to continue to expand in a process called affinity maturation. The staircase then splits; some B cells go on to produce massive amounts of antibody, and some become memory cells, poised to expand when antigen is encountered again later in life. Illustration by Kate Baldwin.

Next, as a B cell clone multiplies, mutations take place in the B cells that increase the strength of receptor binding to the antigen. This is the next rise in the staircase. Then, those clones that bind more strongly to the antigen expand in a process called affinity maturation; this is the second run in the staircase. After that, the staircase splits. Some B cells go on to secrete antibody at a high rate (up to 2,000 molecules per second for several days). Because of the exponential amplification of B cells, and the high rate of antibody production, the body can mount a vigorous battle against a rapidly multiplying invader in the course of a week or so.

Importantly, other B cells become "memory" cells that can persist for many years, poised to respond even more rapidly and forcefully when the antigen is encountered for a second time (Figure 6.3). This is why we usually don't get infected by the same microbe a second time.

To be able to thwart any potential foe, the body needs an army of B cells that make antibodies to different antigens. That poses the big mystery: How can the immune system make millions of different B cells that each make a different antibody?

THE ARSENAL OF IMMUNITY:
SO MANY FROM SO FEW

One important feature to appreciate is how antibodies bind to antigens.

Antibodies are Y-shaped molecules composed of four protein chains: two longer "heavy" and two shorter "light" chains (Figure 6.4). The Y shape emerges from how the chains are put together. The heavy and light chains bind to one another, and the heavy chains themselves are also held together. Where the heavy and light chains combine together, they form a pocket, which is the site where antigen is bound. There are two such antigen-binding sites per individual antibody molecule (one on each arm of the Y). The specificity of antibodies is determined by the sequence of amino acids in the antigen-binding site. Different antibodies have different amino acid sequences in their antigen-binding sites, so the antibody diversity puzzle boils down to how the immune system can generate millions of different sequences of antigen-binding sites.

Antibodies are proteins, so they are encoded by genes. The key, long-elusive mystery was whether the genome contains millions of antibody genes (i.e., millions of heavy and light chain genes), or whether some events took place

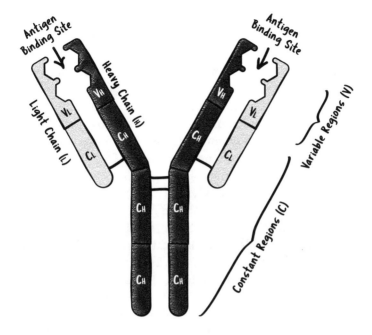

FIGURE 6.4. Antibody structure

Antibodies are made up of four protein chains: two heavy (H) chains and two light (L) chains joined to form Y-shaped molecule. At each end of the Y, the two chains form a pocket, which is the antigen-binding site. These two parts of each chain are more variable in sequence and are called the variable (V) regions; the rest of each chain is essentially invariant and called the constant (C) region. The variable regions are generated by combining gene segments. Illustration by Kate Baldwin.

during the development of B cells that generated diverse antibodies from a smaller number of heavy and light chain genes.

Once gene cloning and DNA sequencing became possible, knowledge of the structure of antibody genes became possible. The breakthrough was made in 1976 by Susumu Tonegawa who discovered that antibody genes were assembled by rearranging pieces of DNA in developing B cells. Tonegawa later received the Nobel Prize for his insight into how a small number of gene pieces could be combined in different ways to generate a very large number of different antibodies.

Light chains, for example, are assembled from three gene segments called V, J, and C. Heavy chain genes, which are located on a different chromosome than light chain genes, are assembled from four gene segments also called V, J, and C, and a fourth D segment. Only one functional light chain gene and one functional heavy chain gene are assembled in any one B cell, so each B cell is a genetic clone that makes one single form of antibody.

The body's total potential arsenal of antibody genes is a fairly straightforward calculation, similar to calculating the number of possible hands in a card game (or the number of chromosome combinations in a sperm or egg). Given the number of cards in each suit, and the number of suits, one can calculate how many different three-, four-, or five-card hands one could draw at random. The same follows for antibody genes. By knowing how many V, J, or D gene segments there are, and that one of each type of gene segment is combined at random, we can calculate the number of possible heavy and light chains. (The C regions are important for other antibody functions, but do not contribute to antigen binding, so they are not part of the diversity calculation).

I will break down the math, but if your eyes start glazing over, just think about shuffling cards and how many hands one can make at random from two, three, or more cards.

For example, humans have 51 heavy chain V segments, 27 heavy chain D segments, and 6 heavy chain J segments. Therefore, if each one of these segments is combined at random with the other segments to assemble parts of heavy chains, humans can make $51 \times 27 \times 6 = 8{,}262$ different heavy chains. That is a lot of heavy chains from just 84 ($51 + 27 + 6$) gene segments.

Humans also have 40 V segments and 5 J segments for one type of light chain, and 30 V segments and 4 J segments for a second type of light chain. Again, if these are assembled at random, then there are 40 × 5 = 200 different light chains of the first type, and 30 × 4 = 120 different light chains of the second type. That makes a total of 320 light chains from 79 gene segments.

One more calculation gets us toward the total number of possible antibodies. It is very similar to the calculation of the number of possible babies one couple can make because, like the combination of sperm and egg, it is the combination of individual light chain and individual heavy chain proteins that creates antibodies and their antigen-binding sites. Which individual light chain and which individual heavy chain are made by a given B cell is a matter of chance, so that if all possible combinations are used, then there are 320 (light) × 8,262 (heavy), or more than 2.6 million different possible human antibodies. All of that firepower comes from just 163 gene segments (84 heavy and 79 light gene segments). That means the body can make more than 10,000 times as many antibodies as it has gene segments in the genome.

For comparison, there are almost 2.6 million possible five-card poker hands that can be drawn from a standard 52-card deck.

But it turns out this impressive number is actually an underestimate of the number of possible antibodies. The mechanism that assembles gene segments has a bit of play in it so that the junctions between gene segments are not precise. This introduces additional DNA sequence variation at the junctions that further increases antibody diversity many-fold. The human body produces about one billion new B cells *every day*, which is enough to ensure that virtually every possible antibody is represented.

Once an antigen enters the picture, however, activated B cells have one more genetic trick. The variable regions of heavy and light chain DNA sequences undergo further mutations at a rate that is about one million times greater than the background rate of other DNA sequences. Called "somatic hypermutation," this is the second rise in the staircase (Figure 6.3). This process further expands the arsenal of antibody diversity at least another ten-fold.

Three random mechanisms, then, generate antibody diversity: the shuffling and joining of gene segments; the independent combination of light and heavy chains; and somatic hypermutation. Altogether, it is estimated that humans can make at least 10 *billion* different antibodies.

The staircase of self-defense illustrates why we vaccinate. It takes days to begin to make antibodies and longer to establish memory. By exposing people to non-living antigens from microbes and establishing memory cells, vaccinated people are two steps ahead of any future exposure to the pathogen, which is why they either don't get the infection or experience a milder case. The staircase also shows why we sometimes deliberately induce animals to make antibodies—in case we need to have antibodies to something *immediately* in an emergency, say for a snakebite.

After Jamie Coots' death, his twenty-one-year-old son Cody assumed pastorship of the church and continued the tradition of snake-handling. Four years later, Cody was handling a timber rattlesnake during a service when the snake struck him above his right ear, puncturing his temporal artery. Bleeding profusely, he gamely tried to continue holding the snake and preaching but had to be carried out of the church. Struggling to breathe, he asked to go to a nearby mountaintop so as to let God decide whether he should live or die.

A member of his congregation thought otherwise and took him to a nearby hospital where the emergency room staff called a Code Blue. They opened Cody's airway and then had him airlifted to a Tennessee medical center, unsure whether he would survive. There, he was put on life support and administered antivenom (made of animal antibodies to snake venom toxins) to try to neutralize and clear the venom. After a ten-day struggle in the intensive care unit, Cody pulled through.

Just one week after he got out of the hospital, he was back to handling snakes.

Sure, he's pushing his luck, but as I'll show in the next chapter, so are we all.

CHAPTER 7

A SERIES OF UNFORTUNATE EVENTS

"Everybody will die, but very few people want
to be reminded of that fact."

—LEMONY SNICKET, *THE AUSTERE ACADEMY*

ONE SPRING DAY IN 1972, it was raining gently on the Blue
Ridge mountains in central Virginia, part of the Appalachian
range. Veteran park ranger Roy Sullivan was manning the
registration station at the Loft Mountain camping area in
Shenandoah National Park when all of a sudden there was
a deafening clap.

"The loudest thing I ever heard," Sullivan later told a local
reporter. "The fire was bouncing around inside the station,
and when my ears stopped ringing, I heard something siz-
zling. It was my hair on fire. The flames were up six inches."
Sullivan snuffed out the flames with his jacket and rushed
to the restroom to cool his burnt scalp.

Sullivan was lucky not to be more seriously injured, or worse. About ten percent of victims of lightning strikes are killed, as bolts may deliver more than 100 million volts. But what made Sullivan even more fortunate was that this was not his first lightning strike, or his second, or even his third. It was the fourth bolt of his career, and his third in a span of four years, all of which occurred within the park.

The first happened thirty years earlier when Sullivan was trying to run away from a fire tower during a heavy thunderstorm. The bolt burned a half-inch strip down his right leg and knocked the nail off his big toe. The second strike was not until 27 years later when he was driving a park truck along Shenandoah's famous, scenic Skyline Drive. That bolt went through the open cab, burned off his eyelashes, eyebrows, and much of his hair, scorched his wristwatch and knocked Sullivan unconscious. His truck rolled to a stop at the edge of a cliff. The third strike was just a year later when a bolt ricocheted off a transformer and knocked Sullivan down in his garden.

At the time of his fourth strike, only two men were known to have been hit three times; in each case the third strike was fatal. With Sullivan's four incidents verified by Park authorities or doctors, the 60-year-old ranger landed in the *Guinness Book of World Records* as the only living person to have been struck four times, earning the monikers "the human lightning conductor," "Lightning Man," or the "Spark Ranger."

Sullivan was often asked, "Why him?" He attributed his "luck" to Providence but admitted he did not understand why God had singled him out so often. Nor, apparently was God satisfied because Sullivan was hit twice more—in 1973 and 1976, before he retired from the Park Service. Then, for good measure, Sullivan was zapped and burned again in 1977 while fishing.

Another possible explanation for Sullivan's record is simply that lightning is an occupational hazard. All sorts of jobs carry a greater chance of some danger: Boxers get hit by fists, preachers get hit by rattlesnakes, park rangers get hit by lightning.

Heck, it turns out that just being alive is an occupational hazard. Look at the graph below:

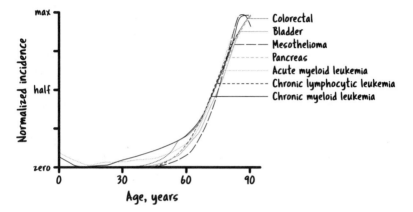

FIGURE 7.1. The incidence of cancer rises 100-fold with age

Illustration by Kate Baldwin based on data from Rozhok and DeGregori (2019).

The curves reflect how the incidence of cancer increases about 100-fold between the ages of thirty and seventy-five; about two of every five people will be struck in their lifetimes.

This pattern was first noticed almost seventy years ago. At the same time, epidemiologists were gathering the first strong evidence of a link between cancer and certain life-style habits or jobs. Researchers have been trying ever since to figure out why and how cancer forms. The increase in the probability of cancer with age or certain activities, coupled with the fact that not everyone will develop the disease, suggested to early researchers that there was a strong element of chance.

In this chapter we will ask: To what degree is cancer a matter of bad genes, bad habits, or bad luck?

"Hooray! I was waiting for him to get to cancer," is probably not your first reaction. Sure, it is not a cheery subject, but we will see exactly why cancer is part of our chance-driven lives—how one of the same processes that brings us fortune can bring us misfortune. I've drawn a picture of the games that go on in our gonads and the lottery at life's conception. Well, in virtually all of the 37 trillion other cells in your body, there is another game of chance unfolding. And one of the cells in your body right now has a good chance of killing you some day. I figured you would want to know about it.

Having made it this far in the story of chance, you are superbly well-prepared to appreciate how some now familiar phenomena—random mutation and selection—create yet another staircase—of cancer. And some of what you learn just might help keep you or a loved one off that staircase.

THE STAIRCASE OF CANCER

It was not just the fact that cancer increases with age that captured epidemiologists' attention. The shapes of the cancer incidence curves are exponential, meaning that the number of cancers rises at a faster rate than the number of years of age. To figure out that rate and what it might signify, epidemiologists applied the inverse of exponentials, called logarithms. When these early researchers plotted the logarithm of death rates from various cancers versus the logarithm of age, they discovered a remarkably consistent mathematical relationship among several different cancers: the death rate from each increased as a function of the sixth power of age (Figure 7.2).

Why six, and what might that mean?

If cancer was a "one-hit" phenomenon, like lightning or snakebite, one might expect strikes to be equally common across different ages. But because cancer increased propor-

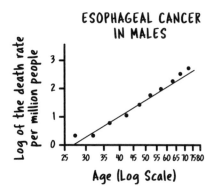

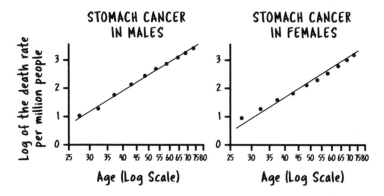

FIGURE 7.2. Cancer is a function of the sixth power of age

The incidence of various kinds of cancer rises as a sixth power of age, which led early researchers to suggest that the genesis of cancer involved a series of mutations. Illustration by Kate Baldwin based on data from Armitage and Doll (1954).

tionally with the sixth power of age, that suggested a cumulative risk and a cumulative, multi-hit process. Even though this relationship was observed before the structure of DNA was solved and the nature of mutation was revealed, a few pioneering epidemiologists boldly suggested that cancer was the result of a series of several successive mutations within a cell that accumulate with age. Even more boldly, based on the discovery that many cancers were a function of the sixth power of age, they proposed that those cancers might specifically require six or seven successive mutations.

In this scenario, researchers pictured an iterative process: an initial mutation occurs in a single cell, and the cell multiplies; later, a second mutation occurs within one of these cells, followed by their multiplication; then a third mutation occurs, and so on. They recognized that if each mutation provided some growth advantage over normal cells, they would outgrow and eventually overgrow normal tissue, i.e., behave as a cancer. Over the ensuing decades, studies of human and animal tumors revealed that cells from the same tumor often shared some obvious chromosomal abnormality, which lent support to the idea that tumors originated from an initial mutation in a single cell. That means that tumors are genetic clones.

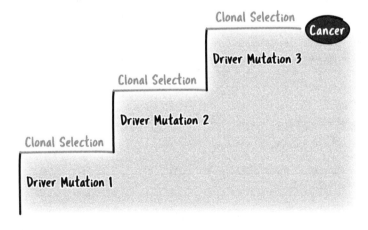

FIGURE 7.3. The Staircase of Cancer

Cancer is a multi-hit process. An initial mutation in a driver gene (the first rise in the staircase) may provide a slight selective advantage to a clone of cells relative to other cells (the run of each stair); a second driver mutation in a cell of the initial clone (mutation 2) may provide a further selective advantage (the second run), and a third driver mutation within a cell of this second clone (mutation 3) may further increase the selective advantage to the point that cancer forms. Illustration by Kate Baldwin.

What this also means is that cancer is another staircase, with new mutations providing each step up (the rise in the stair), and selection providing each step forward (the run of the stair) by propagating the clone containing each new mutation before the next step up occurs (Figure 7.3).

This staircase, though, is merely a conceptual outline of cancer formation. What scientists and physicians desperately want to know is the identity of mutations that initiate each step, and the number of steps in any cancer.

DRIVERS, ACCELERATORS, AND BRAKES

It was not until two decades after the foundational epidemiological studies that researchers started to get a glimpse into the genetics of cancer. One of the major open questions was whether cancer was the result of some general abnormalities, such as differences in chromosome number, or due to the alteration of specific genes. The first breakthroughs came from the identification of very specific chromosome rearrangements that were observed to occur independently in patients with the same type of cancer. These suggested that those cancers at least were due to alterations of specific genes.

Initially, no one knew how many different nor what kinds of genes could contribute to cancer. There are about 20,000 genes in the human genome. If a large fraction of those genes contributed to cancer formation, it would be a chaotic situation to untangle. Eventually, as gene cloning and DNA sequencing became possible, a catalog of gene mutations associated with human cancers began to be assembled. Over several decades of research, about 150 or so genes have been identified that are frequently mutated in cancers. That number indicates that only a small subset (less than 1 percent) of all genes is critical to cancer formation.

The mutated genes fall into two broad functional categories. Mutations in certain genes increase or alter the activity of the proteins they make and promote the formation of cancer; these genes are dubbed oncogenes. In other cancers, certain genes are deleted or inactivated by mutations, which suggests that the normal function of the proteins they make is to check cell growth; these genes are dubbed tumor suppressors.

Another way to think of these two categories is to imagine a car speeding out of control. There are two ways that could happen—a stuck accelerator, or broken brakes. The oncogene mutations cause stuck accelerators, the tumor suppressor mutations cut the brakes. Both categories of mutations drive the formation of cancer so they are referred to in general as "driver" gene mutations.

It does not take much extra speed to create a problem. If a driver mutation provides only a very small selective growth advantage relative to other cells, even less than one percent, this slight advantage compounded day after day, week after week, for many years can create a mass of billions of cells.

Ideally, it would be possible someday to inspect any tumor for all potential driver mutations and to correlate mutations with tumor behavior. Far sooner than expected, that day is already here, and the vista is both enlightening and sobering.

THE SCENE OF THE ACCIDENT

The revolution in the genetic analysis of cancer was catalyzed by dramatic decreases in the time and cost of DNA sequencing, from ~$100,000 per patient a decade or so ago to $1,000 or less today. National and international efforts have created "atlases" of DNA sequences from tens of thousands of cancers.

The task for researchers and clinicians is to comb through the DNA sequence of a cancer to spot driver mutations. The status of each driver gene is assessed by examining the integrity of their sequences. By analyzing the sequences of many cancers, specific associations between certain driver mutations and certain cancers have emerged. For example, certain leukemias almost always involve mutations within one particular driver gene, as do colorectal cancers and pediatric retinoblastomas. Other driver mutations occur in a large fraction of a wide variety of cancers, indicating that they contribute in some general way to cancer formation. And typically, more than one driver mutation is present; most tumors contain two to eight driver mutations—they often have both stuck accelerators and broken brakes.

The multiple driver gene mutations are strong evidence that cancer is generally a multi-hit process. But that is not the only information we can glean from cancer genomes. Driver mutations aren't the only mutations in any given cancer; in fact they are usually the small minority of mutations.

Look at Figure 7.4. It plots the total number of protein-altering mutations found in a wide variety of cancers from fewest to most.

Overall, the number of non-silent mutations in cancers ranges from fewer than 10 to more than 200. Most mutations occur outside of driver genes, exhibit a largely random distribution, and are not implicated in cancer formation. These are dubbed "passenger" mutations because they are merely riding along in a cancer. It turns out the number of passenger mutations is especially informative about the genesis of cancer.

Let's walk through the figure from the left. Note that the number of mutations in pediatric tumors is the fewest of all cancers. The average number of mutations in most adult solid tumors generally spans from around 15 to 75, with a couple kinds of tumors exceeding 125 mutations.

Before I explain more, look at the tumor types along the bottom of the chart. Take a moment to think about why certain tumors fall where they do.

Now let's make some specific comparisons.

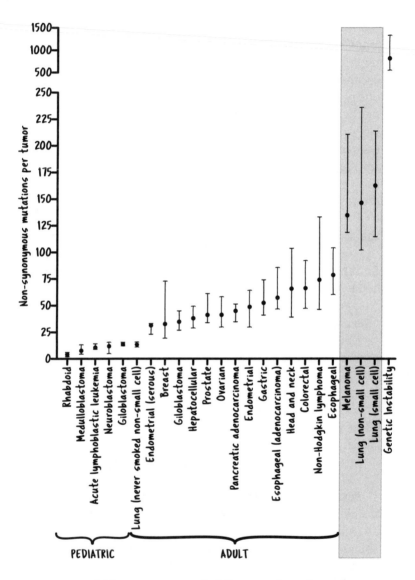

FIGURE 7.4. Different cancers contain different numbers of mutations
Illustration by Kate Baldwin based on data from Vogelstein et al. (2013).

Start with the adult versus pediatric pattern. Why might adult tumors carry more passenger mutations? The most straightforward explanation is that cancers that arise in adult cells have been through more rounds of DNA replication than cancers in children and have thus accumulated more mutations.

But what about differences between adult cancers? Look at the two ends of the adult tumor range. At each end are mutations found in lung cancers, but the lower number (<15) occurs in people who never smoked, while ten times as many mutations occur in tumors from people who smoked (in two distinct types of lung cancer called small cell and non-small cell cancer). What might explain this difference?

Cigarette smoke contains dozens of chemicals that modify or cause damage to DNA. The larger average number of passenger mutations in lung cancers of those who smoke relative to those who don't reveals that smoking increases the frequency of mutations. Indeed, lung cancers contain more mutations than all other cancers in Figure 7.4 except one.

And the identity of that one other cancer is also illuminating; it is melanoma. Skin cancers are correlated with greater sun exposure; the highest rates of skin cancer occur in Caucasians in sunny climates (e.g., South Africa, Australia, and Arizona). The sun produces ultraviolet radiation, and UV rays are well documented to cause DNA damage and mutation. Thus, the larger number of passenger mutations in melanomas reflects that, like smoker's lung cells, these cells have had much greater long-term exposure to mutagens.

Indeed, now that we have the power to rapidly analyze DNA from single cells, researchers have used this capability to see if they can detect driver mutations in normal tissue, to trace the early local events in cancer formation. Analyzing normal-appearing skin tissue from the eyelids of plastic surgery patients, a British team discovered that the cells contained a very large number of mutations with about one-quarter bearing a potential driver mutation.

The larger numbers of mutations in lung cancers, melanoma, and even normal-appearing skin tissue are vivid evidence of why smoking and sun exposure increase the risk of cancer. Mutations occur at random throughout the genome, but the more mutations that occur in a cell's DNA, the greater the chances that a driver gene will get hit, and then a second one, and so on.

The relationship between the number of mutations and the likelihood of cancer is also vividly underscored by one special category of driver mutations. Some cancers contain enormous numbers of mutations (500–1,500) that are literally off the chart in Figure 7.4. It turns out these tumors have mutations in genes that encode proteins that survey and repair DNA damage. Consequently, these tumors' overall mutation rates are much higher, and they are even more likely to acquire additional driver mutations.

We can now appreciate why the risk of cancer increases with age: the more divisions a cell goes through, the more mutations it will accumulate. But cancer is a matter of chance, not certainty. It is a matter of where in DNA random mutations happen to occur. Most cells will not get a driver mutation.

FIGHTING CHANCE

What does this new knowledge tell us about the cause of cancer? Well, it depends on whom you ask. A geneticist will say that driver mutations cause cancer, and that is true, but not necessarily that helpful for how we live our lives and look after our loved ones. The more pertinent and pressing question is what causes those driver mutations? I suggest there are three parts to the answer to that question.

The first is underscored by the pediatric tumors in Figure 7.4. They carry a small number of mutations that just happen to have hit driver genes. There is no lifestyle component to consider; they are the result of terribly bad luck and thankfully relatively rare. One of the kind things you can do for anyone whose child is struck is to assure them that there was nothing they did to cause it, and nothing they could have done to prevent it.

The second part of the answer comes from lung, skin, and cervical cancers. These cancers also include an element of bad luck, but there are clearly ways to stay off those staircases. For example, smoking increases the lifetime risk of lung cancer ten- to twenty-fold. As Kurt Vonnegut put it, "The public health authorities never mention the main reason many Americans have for smoking heavily, which is that smoking is a fairly sure, fairly honorable form of suicide." Similarly, more people are diagnosed every year with skin cancer than all other cancers combined, but a variety of studies have found that regular use of UV-reducing sunscreen in at-risk populations can reduce its incidence. We should all heed the advice Vonnegut once offered to new college graduates, "Use sunscreen! Don't smoke cigarettes." Likewise, the vast majority of cervical cancers (as well as many oral, head, and neck cancers) are caused by infection with human papilloma virus (HPV), which is now largely preventable by vaccination (if you believe in that sort of thing!).

The third part to the answer comes from the majority of the adult tumors on the chart. The number of mutations in these cancers reflects how mutation is inexorable, an inevitable byproduct of being alive and copying DNA. While some proportion are influenced by lifestyle and environmental exposure, most adult cancers, too, are largely a matter of bad luck—the result of a series of unfortunate events— but also a consequence of the good fortune of living longer lives.

But the most important question is, what does all of this mean for fighting cancer once it occurs? The very good news is that this new knowledge of chance mutations has brought new power and new hope. The big-brained ape sculpted by the chaos of the Ice Age is using that oversized organ to battle chance.

Twenty-five years ago, we did not know where to look in most cancers for driver mutations, and even if we did, we were largely powerless. But beginning in 1998, a new class of drugs was invented that target specific molecules produced by driver mutations. Today, there are scores of drugs that target molecules involved in the growth and spread of more than thirty kinds of cancer, and many more candidates are under investigation. So, if and when lightning strikes, we now have an increasingly better chance of survival.

A CONVERSATION ABOUT CHANCE

"Tiger got to hunt, bird got to fly;
Man got to sit and wonder 'why, why, why?'
Tiger got to sleep, bird got to land;
Man got to tell himself he understand."

—KURT VONNEGUT, *CAT'S CRADLE*
(THE BOOKS OF BOKONON)

IN HIS QUASI-AUTOBIOGRAPHICAL novel *Slapstick*, Kurt Vonnegut tells the story of the death of his sister Alice from cancer at age forty-one. He and his brother Bernard visited her on what would be the last day of her life.

"(H)ers would have been an unremarkable death statistically, if it were not for one detail, which is this: Her healthy husband, James Carmalt Adams, the editor of a trade journal for purchasing agents, which he put together in a cubicle on Wall Street, had died two mornings before—on 'The Brokers'

Special,' the only train in American railroading history to hurl itself off an open drawbridge.

"Think of that.

"This really happened."

The great humorist manufactured many fantastic scenarios in his tales, but the train wreck really did happen just as he said. On the morning of September 15, 1958, a New Jersey Central commuter train carrying about 100 passengers toward Jersey City somehow drove slowly through three signal lights and an automatic derailer and carried on 550 feet before plunging 50 feet off the open drawbridge. Two coaches of the five-coach train hit the water while a third dangled mid-air for more than two hours. Despite the heroic efforts of rescue crews, forty-eight people died, including Vonnegut's brother-in-law.

Given his sister's grave condition, and her concern for her four young children who were about to be in the sole care of her husband, Vonnegut and his brother elected not to tell her about the tragedy. She found out anyway when a fellow patient gave her a copy of the newspaper and she saw her husband among the list of the dead or missing. Vonnegut describes her reaction, and his own:

"Since Alice had never received any religious instruction, and since she had led a blameless life, she never thought of her awful luck as being anything but accidents in a very busy place.

"Good for her."

Accidents in a very busy place, indeed. I have shown how we know that we are all here, both collectively and individually, through a series of accidents—cosmological, geological, and biological accidents. I have also shown how and why some of us will depart via accident.

Our chance-driven world is a profound revelation. It is astonishing that blind chance is the source of all novelty, diversity, and beauty in the biosphere. I hope that you are wonderstruck at what an asteroid, sliding tectonic plates, and a fibrillating polymer of just four bases have wrought.

But our chance-driven existence also poses the unsettling quandary that we don't live in the best of all possible worlds, but a world as novelist Christian Jungerson put it of "unmerciful randomness, immense chaos, and constant biological vulnerability." This view, of course, shatters traditional beliefs about humanity's place in the larger scheme.

As Monod's critics decried, chance puts God out of a job, or at least many of the jobs we have traditionally assigned to Him or Her. God is not in the conception business choosing the winning sperm and egg, nor the genetic engineering business designing creatures' DNA and traits, nor the weather-making business, nor the cancer business, nor the pandemic business.

One recourse in the face of such claims is simply to deny chance. But should we have the courage to accept the pervasive role of chance, some challenging questions arise about the meaning and purpose of our lives: If we are here by accident, not by Design, what are we supposed to do? How might we live in the face of this knowledge?

My first impulse is to pull a Dr. McCoy ("Bones" from *Star Trek*) and plead, "Dammit, Jim, I'm a scientist, not a philosopher." I think that these questions are for each of us to decide for ourselves. But because this is the Afterword and there may be some expectation for my thoughts on the matter, I'll offer some possible replies with the help of some of my favorite thinkers.

Our accident-driven world and humans' struggle for meaning were recurrent themes throughout Vonnegut's works. Right after his sister Alice's death, Vonnegut wrote his second novel *The Sirens of Titan* (1959), which was set in the future and began:

"Everyone now knows how to find the meaning of life within himself.

"But mankind wasn't always so lucky. Less than a century ago men and women did not have such easy access to the puzzle boxes within them.

"They could not name even one of the fifty-three portals to the soul.

"Gimcrack religions were big business."

One of the two main characters in the novel (Winston Rumfoord) establishes the Church of God the Utterly Indifferent whose principal teachings are: "Puny man can do nothing at all to help or please God Almighty, and Luck is not the hand of God." After the second main character (Malachi Constant) is abducted from and later returns to Earth, he declares one of the central themes of the Church: "I was a victim of a series of accidents, as are we all."

Vonnegut's books helped me to realize that, next to scientists, the one group of people that seem least inclined to think everything happens for a reason, and that blind chance governs the world, are humorists and comedians. So many great present-day comedians—Seth MacFarlane, Eric Idle, Bill Maher, Ricky Gervais, Sarah Silverman, Bill Burr, Eddie Izzard, Lewis Black, and more, and late greats from Mark Twain to Vonnegut to George Carlin have rejected traditional beliefs about humanity's place and purpose while making us laugh at the absurdity of some of those beliefs.

So many very funny people, in fact, that it made me wonder, why is this so? What do scientists and comedians have in common? Why are comedians drawn toward such subjects? How do they think we should live in the face of the ubiquity of chance? I decided to reach out and ask some of them directly.

Then it occurred to me: Wouldn't it be fun to bring them together, not just with each other but also with some of my literary and scientific heroes for a conversation about chance and all of its implications? I began to imagine sitting in a living room with all of these funny and clever people while they let loose.

Logistically, that was impossible because most are all terrifically busy writing, acting, and touring, and a few are quite inconveniently dead. I decided the best way to share their collective wit and wisdom was to create a script based on what they told me or have said to others. What follows is how the conversation *might have* gone (most quotes are verbatim; see Notes).

A CONVERSATION ABOUT CHANCE

CAST OF CHARACTERS

Albert Camus	Writer-philosopher, Nobel Prize for Literature 1957
Ricky Gervais	Comedian-actor, creator of *The Office*, *After Life*
Eric Idle	Founding member of Monty Python, songwriter
Eddie Izzard	Comedian-actor, specials include *Force Majeure*
Bill Maher	Comedian, host of *Real Time with Bill Maher*
Seth MacFarlane	Comedian-animator-actor, creator of *The Family Guy*
Jacques Monod	Nobel biologist (1965), wrote *Chance and Necessity*
Sarah Silverman	Comedian-actress, specials include *Speck of Dust*
Kurt Vonnegut	Author of *The Sirens of Titan*, *Slaughterhouse-Five*
Sean Carroll	Moderator, no Nobel Prize or comedy special

CARROLL

Thank you all for being here. Let's get started by talking about chance, accidents, the "What ifs?" in your lives. Seth, I started this book off with the story of your very close call on 9/11. Has that experience affected your life?

MACFARLANE

Not especially. I'm not a fatalist. I'm not a religious person. I'm sure there are close calls that we're not even aware of hundreds of times a year. You cross the street, and if you'd crossed the street two minutes later, you'd have been hit by a car, but you'd never know it. I'm sure that kind of stuff happens all the time.

SILVERMAN

I am insanely lucky to be alive. I had a freak case of epiglottitis in 2016. I didn't know going in, because I was instantly put on medication. But I learned later I was supposed to die.

CARROLL

What happened?

SILVERMAN

Don't even know why I went to the doctor, it was just a sore throat. He looks down my throat, and he's like: "We're going to the emergency room." I had an abscess at the top of my windpipe, which is where your breathing comes from. This abscess was either gonna grow another millimeter and stop my breathing and I would die, or it was gonna explode, and it's filled with poison, and that would kill me. When I woke up 5 days later I didn't remember anything.

CARROLL

How has this experience affected you?

SILVERMAN

I spent the first two days home kind of free-falling from the meds/lack of meds and the paralyzing realization that nothing matters. Luckily that was followed by the motivating revelation that nothing matters. All of this has maybe made me a little more . . . I don't know if grateful is the word because it sounds a little cliché.

MONOD

I am also grateful. I turned down an Arctic voyage on a boat that ended up sinking. The ship was called the *Pourquoi-Pas?* She was commanded by a famous captain and explorer. I sailed on it to Greenland as a naturalist in 1934. Two years later, they invited me to go again. At the last minute, I decided instead to go to California. The ship sank in a hurricane off Iceland and all but one aboard died.

CARROLL

And what about later during the war? Both you and Albert were in the Resistance. That was very risky.

CAMUS

Well, yes. It was quite dangerous. But I was lucky.

CARROLL

How so?

CAMUS

I did not get caught, but many of my comrades were captured or killed. The woman who recruited me was arrested on a day she was supposed to meet me, and was deported to a camp. I was once stopped in the street and searched, but they did not find the papers I had, or I would have joined her, or worse.

MONOD

The Gestapo was everywhere in Paris, we never knew if the next rendezvous would be our last. I remember thinking as I would approach a meeting at a safehouse, "Do I turn in and risk never seeing my family again? Or do I walk past and go home." Three days before D-Day they caught almost all of the commanders in Paris. I was lucky to be out of town.

VONNEGUT

I got caught. I was an Army Scout and taken prisoner by the Germans during the Battle of the Bulge. We were in this gully about as deep as a World War I trench. There was snow all around. Somebody said we were probably in Luxembourg. We were out of food. The Germans could see us, because they were talking to us through a loudspeaker. They told us our situation was hopeless, and so on.

They sent in eighty-eight millimeter shells. The shells burst in the treetops right over us. Those were very loud bangs right over our heads. We were showered with splintered steel. Some people got hit. Then the Germans told us again to come out. So we did.

They said the war was all over for us, that we were lucky, that we could now be sure we would live through the war, which was more than they could be sure of.

CARROLL

Were you lucky?

VONNEGUT

Yes and no. Six weeks later, I was a POW in Dresden when it was firebombed by the Allies. We never expected to get it. There were very few air-raid shelters in town and no war industries, just cigarette factories, hospitals, clarinet factories. Then a siren went off—it was February 13, 1945—and we went down two stories under the pavement into a big meat locker. When we came up the city was gone.

Every day we walked into the city and dug into basements and shelters to get the corpses out, as a sanitary measure. It was a terribly elaborate Easter-egg hunt. Twenty-five years later I wrote *Slaughterhouse-Five*.

IDLE

Here is another true story that could have been taken right from one of Kurt's novels. My father was killed hitchhiking home from World War II.

VONNEGUT

Christ.

IDLE

He served in the Royal Air Force from 1941 to the end of World War II as a rear-gunner/wireless operator in a Wellington bomber, the most dangerous seat in the plane, and emerged unscathed. At Christmas 1945, seven months after the end of the war in Europe, the trains were all full so he hitched a lift home in the back of a lorry carrying a load of steel. A car swerved to avoid traffic, the truck veered off the road, the load of steel shifted and crushed him. He died in hospital on Christmas Eve. I was three years old.

CAMUS

Terrible. My father, too, was killed—in World War I in the Battle of the Marne [1914]. I was not yet a year old.

GERVAIS

So sorry, mates. War shaped my fate, too, but in the opposite direction, my Mum and Dad met in England during a blackout.

IDLE

How romantic!

GERVAIS

But it was a long time before I came around. I am much younger than my three siblings. I remember saying to my mum, when I was about 11 or 12, 'Why are the others so much older than me?' And she went, 'Because you were a mistake.'

VONNEGUT

As are we all.

IZZARD

World War II is it, World War II. That's the time He should have come down. If He made us all, and we were praying like crazy, and fifty million dying, and this idiot with the mustache, you know, that was the time to come down.

That didn't happen, so I don't think He's coming. He doesn't come for tsunamis, He doesn't come for earthquakes, He doesn't come for world wars. He's not coming.

CARROLL

Eddie is cutting to the heart of the matter, isn't he? It really doesn't look like anyone is in charge of human affairs, watching out for us.

Which brings me to a key question for this group: Why do you think so many comedians conclude that there is no God?

IDLE

It saves time.

MAHER

We are talking about humanistic gods people pray to, that they think can intervene in our lives, who run sort of a heaven-and-hell operation for the afterworld. That sort of traditional religion is what we're talking about.

IZZARD

Religion is just full of stories. They're stories, in which case Lord of the Rings is the same.

MAHER

Right. We are used to the story of a man living inside of a whale for three days, we're used to the idea that a space god impregnated a virgin and had a child who was really him, and sent him on a suicide mission to earth, which he survived. If Christianity were the new religion, we would consider it just as crazy as Scientology.

IDLE

I think we [Monty Python] should now apply for tax-deductible status on the grounds that we are funnier than Scientology . . . we could be a religion. Briantology.

CARROLL

Fine to think that, but why go public, why bring religion into your acts? Isn't that risky for you? Kurt's books are among those banned most often. *Family Guy* draws fierce protests. *Life of Brian* was picketed and banned in various places.

IDLE

Comedy is telling the truth. It's the Emperor's New Clothes. Everything is on the table. The *Life of Brian* is still playing after forty years . . .

MAHER

When people laugh, somewhere inside they know it might be true.

VONNEGUT

I agree. The telling of jokes is an art of its own, and it always rises from some emotional threat. The best jokes are dangerous, and dangerous because they are in some way truthful.

IDLE

Truth is the point of comedy. It's usually saying the right thing at the wrong time.

CARROLL

A fair question might be what do comedians know about God anyway?

SILVERMAN

I don't know if there is a God. I mean, I cannot imagine there is a God. But I don't know. Neither do you, you know? But if there is a God, it's a God that's, like, totally fine with murder and, uh, children starving and . . . spin class. You know, all the atrocities of life.

IZZARD

Religious people might think it goes on after death. My feeling is that if that is the case it would be nice if just one person came back and let us know it was all fine, all confirmed. Of all the billions of people who have died, if just one of them could come through the clouds and say, you know, "It's me Jeanine, it's brilliant, there's a really good spa," that would be great.

GERVAIS

Since there is nothing to know about god, a comedian knows as much about god as anyone else. An atheist however is alone in knowing that there is nothing to know so probably has the edge. An atheist comedian can make people laugh about belief or lack of it. A good atheist comedian can make people laugh AND think about belief or lack of it.

CARROLL

And humor makes the truth go down easier . . .

MacFARLANE

Same goes for the traditional sci-fi method of storytelling, which is to take elements of our society, whether it be social or political or scientific, and find ways to tell stories about those things in an allegorical fashion through the lens of sci-fi.

VONNEGUT

When Shakespeare figured the audience had had enough of the heavy stuff, he'd let up a little, bring in a clown or a foolish innkeeper or something like that, before he'd become serious again. And trips to other planets, science fiction of an obviously kidding sort, is equivalent to bringing in the clowns every so often to lighten things up.

This is in contrast to a more didactic approach like our friend Camus. How did you phrase the main question in *The Myth of Sisyphus*?

CAMUS

There is but one truly serious philosophical problem, and that is suicide.

VONNEGUT

So there's another barrel of laughs from literature.

CARROLL

Which raises a question I just have to ask Eric. Why do you sing the title song to *The Meaning of Life* in a French accent?

IDLE

Because Philosophy like wine is better in French.

CARROLL

How was that film received in France?

IDLE

It won the Grand Jury Prize in 1983 [at Cannes].

CARROLL

Humor, science fiction, cartoons, songs—maybe scientists have something to learn from all of you about ways to make a point?

MAHER

In a culture that needs caffeine-free cherry chocolate diet Coke, you'd best deliver information with entertainment.

IDLE

Well the point about science is that it can be tested. Comedy is tested on the audience. If they laugh it works.

CARROLL

Many of you are very strong proponents of science.

MACFARLANE

It's like the civil-rights movement. There have to be people who are vocal about the advancement of knowledge over faith.

VONNEGUT

I love science. All humanists do.

We once had a memorial service for Isaac Asimov, and at one point I said, "Isaac is up in Heaven now." That was the funniest thing I could have said to a bunch of Humanists. I rolled them in the aisles. It was several minutes before order could be restored.

GERVAIS

Science seeks the truth. And it does not discriminate. For better or worse it finds things out. . . . It doesn't get offended when new facts come along. It doesn't hold on to medieval practices because they are tradition. If it did, you wouldn't get a shot of penicillin, you'd pop a leach down your trousers and pray.

CARROLL

So, with no one to pray to and no afterlife to seek, by what code should we live our lives?

IZZARD

There is a rule in every major religion, called the Golden Rule. Essentially: treat other people the way you'd like to be treated yourself. If we all did this, the whole world would work instantaneously. Praying, meditation—fine. But just follow the Golden Rule and the whole world works. Making the world work could be that simple.

GERVAIS

"Do unto others . . ." is a good rule of thumb. Forgiveness is probably the greatest virtue there is. But that's exactly what it is—a virtue. Not just a Christian virtue. No one owns being good.

MAHER

No ethicist has a problem with the Golden Rule, of course, but we don't know why it has to be attached to ancient myths and superstitions. It's fabulous on its own; it didn't have to come down via a burning bush.

GERVAIS

And that's where spirituality really lost its way. When it became a stick to beat people with. Do this or you'll burn in hell.
  You won't burn in hell. But be nice anyway.

CARROLL

So to live a meaningful, happy life, how should we spend our time?

VONNEGUT

We are here on Earth to fart around and don't let anyone tell you any different.

CAMUS

Maybe so. But of all the things we might do, the act of creation is the most significant. Creation is both a key to happiness and a gift to the future.

GERVAIS

Yes! You should bring something into the world that wasn't in the world before. It doesn't matter what that is. It doesn't matter if it's a table or a film or gardening—everyone should create. You should do something, then sit back and say, "I did that."

MONOD

We must create both things and ideas. There is a basic misconception about the role of science: that the purpose of science is to create technology, when technology and applications are by-products. The most important results of science have been to change our view of ourselves, the meaning of our existence.

GERVAIS

It always comes back to us—why are we here? Well, we just happened to be here, we couldn't choose it. We're not special, we're just lucky; and this is a holiday. We didn't exist for 14 and a half billion years. Then we got 80 or 90 years if we're lucky, and then we'll never exist again. So we should make the most of it.

CARROLL

And I think you have all said how to make the most of it— Tell the truth, treat others kindly, create, and for God's sake, laugh. Thank you for that, and for all your creations.

ACKNOWLEDGMENTS

AS THE LATE AND TRULY GREAT Tom Petty sang, "Even the losers get lucky sometimes." It is hard to believe sometimes just how much good luck has come my way. I have many thanks to give.

It all begins with family. Thanks to Mom and Dad, who gave me a decent set of chromosomes and three great and always encouraging siblings, Jim, Nan, and Pete. And thanks to my four wonderful sons—Will, Patrick, Chris, and Josh—now all grown and who each reviewed a draft of the book and provided great input. And eternal thanks to my wife Jamie—Petty would say I got lucky babe when I found you.

Huge thanks to Dr. Megan McGlone, with whom I am very fortunate to be able to work on this, our fifth book together. Thanks for the feedback on the story, for curating and assembling the bibliography and notes, for coordinating the art package and securing permissions, for preparing the manuscript, and most of all, for keeping her good humor throughout the project. And congratulations on the doctorate!

Big thanks as well to Kate Baldwin for her creativity and taste in crafting the main figures, and to Natalya Balnova for the chapter sketches and cover art.

Special thanks to Eric Idle for taking the time to answer my questions for the Conversation, and to Alana Gospodnetich for connecting us; to Eddie Izzard for his input, and agent Max Burgos for connecting us; to Seth MacFarlane for his contribution, and Joy Fehily for facilitating; and to Nanette Vonnegut for background on her father. Thanks also to Sarah Silverman, Bill Maher, and Ricky Gervais for their wit and wisdom; I hope that they appreciate how their words have been repurposed.

Many thanks to my colleagues Eric Haag, Antonis Rokas, David Elisco, and Laura Bonetta who carefully reviewed the manuscript and provided detailed feedback, and to two anonymous reviewers for their thoughtful and constructive comments. And thanks to Dr. Caitlin Schrein for guidance on human paleontology and Dr. Hashim Al-Hashimi for answering my questions about his discoveries concerning the "quantum fibrillation" in DNA bases.

Special thanks to Alison Kalett, my editor at Princeton University Press who embraced the book from the get-go and gently suggested many ways to make it better. And many thanks to my agent Russ Galen for his enthusiasm for this project and fifteen years of very wise counsel.

And finally, thanks to Jacques Monod (1910–1976) whom I never got the chance to meet, but who has been a lodestar since I first read *Chance and Necessity* more than forty years ago. Few people, let alone scientists, have lived fuller lives or with greater clarity of purpose.

Jacques Monod in the Red Sea. Photo courtesy of the Archives of the Pasteur Institute, used by permission of Olivier and Philippe Monod.

NOTES

PAGE

vi **All persons, living:** Vonnegut (1998), p. xi.

INTRODUCTION

PAGE

1 **The crowd whooped:** "Tiger Woods' First Ace in 1996." *PGA TOUR*. Video. www.pgatour.com/video/2014/01/21/tiger_s-first-ace.html

2 **Comrade General Kim:** Gallo, DJ. "Kim Jong-Il Was a Dictator Who Went To Great Links." *ESPN*. 19 Dec 2011. www.espn.com/espn/page2/index/ _/id/7369649; Girard, Daniel. "Kim Jong-Il Once Carded 38-under Par at Pyongyang Golf Course." *The Star*. 19 Dec 2011. Toronto Star Newspapers Ltd. www.thestar.com/sports/golf/2011/12/19/kim_jongil_once_ carded_38under_par_at_pyongyang_golf_course.html.

2 **There are only two:** Read more about the real story here—Sens, Josh. "Behind Kim Jong Il's Famous Round of Golf." *GOLF.com*. 1 June 2016. EB Golf Media LLC. www.golf.com/golf--plus/behind-kim-jong-ils-famous-round-golf

2 **However, the odds:** "Hole In One Odds: Golf's Rare Feat." *US Hole In One*. 22 July 2008. www.holeinoneinsurance.com/news/2008/07/hole-in -one-odds-golfs-rare-feat.html

2 **What makes Jong-Il's:** "Pyongyang Golf Course Scorecard & Course Layout." *Asian Senior Masters*. 1 Oct 2005. Asian Seniors Tour. www .asianseniormasters.com/newsdetails1.asp?NewsID=1076

2 **And what's the chance:** "Kim Jong Il: 10 Weird Facts, Propaganda." *CBS News*. 19 Dec 2011. CBS Interactive. www.cbsnews.com/media/kim-jong -il-10-weird-facts-propaganda/5/

3 **We flock to casinos:** Prince, Todd. "Number of Young, First-Time Visitors to Las Vegas on the Rise." *Las Vegas Review-Journal*. 5 Apr 2017. www .reviewjournal.com/business/tourism/number-of-young-first-time -visitors-to-las-vegas-on-the-rise/

3 **That's how the casinos:** Katsilometes, John. "Reports: Britney Spears to Make $500K Per Show in Las Vegas." 2 July 2018. *Las Vegas Review-Journal*. www.reviewjournal.com/entertainment/entertainment-columns/kats/ reports-britney-spears-to-make-500k-per-show-in-las-vegas/

4 **The Casino made:** Huff and Geis (1959), p. 28–29.

4 **Finally, after almost:** Arie, Sophie. "No 53 Puts Italy Out of its Lottery Agony." *The Guardian*. 10 Feb 2005. Guardian News and Media Limited. www.theguardian.com/world/2005/feb/11/italy.sophiearie.

5 **I say "close":** Grech (2002). There is also some evidence that parents with all boys have a slight tendency to produce another boy, not to break the streak: Rodgers and Daughty (2001).

5 **They have also found:** Croson and Sundali (2005); Clark et al. (2009).

6 **"I think she just":** Bleier, Evan. "California Man Wins $650K in Lottery Day After Wife Dies from Heart Attack." *United Press International*. 31 March 2014. www.upi.com/Odd_News/2014/03/31/California-man-wins-650K-in -lottery-day-after-wife-dies-from-heart-attack/4731396267665/

6. **"everything happens":** "Winner Stories: Timothy McDaniel." *calottery*. California State Lottery. www.calottery.com/win/winner%20stories /timothy-mcdaniel

6 **For example, when:** Carroll (2013), p. 332.

7 **"the most important":** J. Monod, interview with Gerald Leach in Paris, January 3, 1967, broadcast on the BBC February 1, 1967, transcript MON. Bio 09, Fonds Monod, SAIP, p. 11.

7 **"[T]he 'secret of life'":** Monod (1971), p. xii.

8 **"We call these":** Monod (1971), p. 112.

8 **"Pure chance":** Monod (1971), p. 112–113.

8 **"Man was the product":** Monod (1969).

9 **"one of the strongest":** Peacocke (1993), p. 117.

9 *Anti-Chance:* Schoffeniels (1976).

9 *Beyond Chance and Necessity:* Lewis (1974).

9 *God, Chance, and Necessity:* Ward (1996).

9 **"It is not necessary":** Sproul (1994), p. 3.

9 **"Chance as a real":** Sproul (1994), p. 214.

10 **Surveys reveal that:** Pew Research Center. "When Americans Say They Believe in God, What Do They Mean?" *Pew Research Center*. 25 Apr 2018. www.pewforum.org/2018/04/25/when-americans-say-they-believe -in-god-what-do-they-mean/

CHAPTER 1

PAGE

15 **"The Age of Reptiles":** Cuppy (1941), p. 104.

16 **A short while later:** McLaughlin, Katie. "MacFarlane: Angry Jon Stewart Call An 'Odd Hollywood Moment.'" *Piers Morgan Tonight*. 6 Oct 2011. Video. CNN Entertainment. www.cnn.com/2011/SHOWBIZ/10/05/seth .macfarlane.pmt/index.htm.

16 **They later flew on:** Weidinger, Patrick. "10 Famous People Who Avoided Death on 9/11." *Listverse*. 12 Dec 2011. listverse.com/2011/12/12/10-famous -people-who-avoided-death-on-911/

16 **"Alcohol is our friend":** McLaughlin, Katie. "MacFarlane: Angry Jon Stewart Call An 'Odd Hollywood Moment.'" *Piers Morgan Tonight*. 6

Oct 2011. Video. CNN Entertainment. www.cnn.com/2011/SHOWBIZ /10/05/seth.macfarlane.pmt/index.htm.

19 **Alvarez realized something:** Alvarez (1997).

21 **Chemical analyses:** Alavarez et al. (1980); Smit and Hertogen (1980).

21 **He figured that:** Alavarez et al. (1980).

23 **These glass spherules:** Smit (1999).

23 **Finally, in 1991:** Hildebrand (1991).

23 **The collision induced:** Schulte et al. (2010).

24 **The immediate effect:** Robertson et al. (2013).

24 **Temperatures on land:** Artemieva et al. (2017); Brugger et al. (2017); Gulick et al. (2019).

24 **The massive injection:** Henehan et al. (2019).

24 **The roll call of victims:** Robertson et al. (2013).

25 **Far richer than:** Vajda and Bercovici (2014).

25 **Many kinds of birds:** Field et al. (2018).

26 **Similarly, there is:** Rehan et al. (2013).

26 **Now consider the fact:** Gallala et al. (2009).

26 **Crocodiles and turtles:** Robertson et al. (2004).

26 **Snakes as a group:** Longrich et al. (2016).

27 **The fossil record:** Field et al. (2018).

27 **That evolutionary "tree":** Jarvis et al. (2014).

28 **Frogs date back 200 million:** Feng et al. (2017).

28 **Both the fossil record and:** O'Leary et al. (2013).

30 **Scores of species:** Longrich et al. (2016).

30 **And third, within:** Lyson et al. (2019).

30 **It turns out that:** Mazrouei et al. (2019).

31 **The rocks around:** Schulte et al. (2010).

31 **Geologists figure that:** Kaiho and Oshima (2017).

CHAPTER 2

PAGE

34 **Jack Blackburn:** Casey, Mike. "Joe Grim: How to Take It and Then Some." *Boxing.com*. 3 Sept 2013. www.boxing.com/joe_grim_how_to_take_it_ and_then_some.html

34 **"I am Joe Grim":** "Grim Stays the Limit with Fitz." *St. John Daily Sun.* 24 Oct 1903. Accessed at: news.google.com/newspapers?nid=37&dat =19031024&id=HwAIAAAAIBAJ&sjid=9zUDAAAAIBAJ&pg=2275, 1041090

34 **"bob up serenely":** "Grim Stays the Limit with Fitz." *St. John Daily Sun.* 24 Oct 1903. Accessed at: news.google.com/newspapers?nid =37&dat=19031024&id=HwAIAAAAIBAJ&sjid=9zUDAAAAIBAJ&pg =2275,1041090

34 **He stayed down:** "Joe Grim Issues Defi to Jeffries." *Deseret News.* 2 Nov 1903. Accessed at: news.google.com/newspapers?nid=Aul-kAQHnToC&dat =19031102&printsec=frontpage&hl=en

36 **Fitzsimmons promptly:** "Grim Stays the Limit with Fitz." *St. John Daily Sun.* 24 Oct 1903. Accessed at: news.google.com/newspapers?nid =37&dat=19031024&id=HwAIAAAAIBAJ&sjid=9zUDAAAAIBAJ&pg =2275,1041090; Casey, Mike. "Joe Grim: How to Take It and Then Some." *Boxing.com.* 3 Sept 2013.

36 **"I don't believe":** Ehrmann, Pete. "'Iron Man' Joe Grim Found Fame with a Thick Skull." *OnMilwaukee.com.* 7 Jan 2013. onmilwaukee.com /sports/articles/joegrim.html.

36 **"The Human Punching Bag":** Ehrmann, Pete. "'Iron Man' Joe Grim Found Fame with a Thick Skull." *OnMilwaukee.com.* 7 Jan 2013. onmilwaukee.com/ history/articles/joegrim.html.

37 **However, newly discovered:** Lyson et al. (2019).

38 **The Eocene-Oligocene transition:** Blois and Hadly (2009).

38 **The end of the Pliocene:** Pimiento et al. (2017).

38 **And the end:** Koch and Barnosky (2006).

39 **So, the potential causes:** Kennett and Stott (1991).

39 **The climate record:** see Figure 1; McInerney and Wing (2011).

40 **Three later cene-changes:** Hren et al. (2013).

40 **a gradual cooling:** Filippelli and Flores (2009).

40 **and a very sharp cooling:** Alley (2000).

41 **It is thought:** Mazrouei et al. (2019).

41 **Impactors of this size:** Hughes (2003).

41 **The only known:** Goff et al. (2012).

42 **Just a few years ago:** Schaller et al. (2016); Schaller and Fung (2018).

42 **The map of the world:** Scotese, Christopher R. *PALEOMAP Project.* www .scotese.com.

42 **Average global surface:** Anagnostou et al. (2016).

42 **tropical forests attained:** Hansen and Sato (2012); Wing et al. (2005).

43 **In the late Eocene:** Carter et al. (2017).

43 **Carbon dioxide is said:** Lacis et al. (2010).

43 **CO2 levels were extremely:** Anagnostou et al. (2016).

44 **Tectonic forces drove:** Bouilhol et al. (2013).

44 **The Indian plate's:** Kumar et al. (2007).

44 **Pioneering geochemist Wally Broecker:** Broecker (2015).

44 **The plate is about 100 km:** Kumar et al. (2007).

47 **Whatever the catalyst(s):** Martínez-Botí et al. (2015).

49 **Over the past:** Barker et al. (2011); Snyder (2016).

50 **Twenty-five times:** Weart (2003); Greenland Ice-Core Project (GRIP) Members (1993); Alley et al. (1993); Mayewski et al. (1993).

51 **They then warmed:** Alley (2000); Fiedel (2011).

51 **The timing between climate:** Ditlevsen et al. (2007); Lohmann and Ditlevsen (2018).

52 **"The paleoclimate record":** Broecker (1995).

52 **Beginning after:** Tierney et al. (2017).

53 **"When I saw her":** Leakey (1974), p. 159–160.

53 **It also remains:** Potts et al. (2018); Behrensmeyer et al. (2018).

54 **At least 16 major:** "Olorgesailie, Kenya." *East African Research Projects: Human Evolution Research.* Smithsonian Institution. humanorigins.si .edu/research/olorgesailie-kenya

54 **These species were:** Potts et al. (2018).

54 **Over this same 200:** Brooks et al. (2018).

55 **Hominids gained control:** Potts (2013); Gowlett (2016).

55 **And it has long:** deMenocal (1995); Potts (2013); Maslin et al. (2014).

CHAPTER 3

PAGE

61 **"O MOST powerful":** "1789 U.S. Book of Common Prayer." *Online Anglican Resources at Justus.anglican.org.* justus.anglican.org/resources /bcp/1789/Prayer_at_Sea_1789.htm.

61 **They also had many:** "Royal Navy Loss List Searchable Database." *Maritime Archaeology Sea Trust.* 4 Feb 2018. www.thisismast.org/research /royal-navy-loss-list-search.html.

62 **"Nothing could be":** King (1839), p. 179.

63 **"full of zeal":** Letter, Francis Beaufort to Robert FitzRoy, September 1, 1831, Darwin Correspondence Project, "Letter no. 113," accessed on 25 March 2019, www.darwinproject.ac.uk/DCP-LETT-113.

63 **"None but those":** Darwin (2001), p. 132.

63 **"We have had":** Letter, Charles Darwin to Emily Catherine Langton, November 8, 1834, Darwin Correspondence Project, "Letter no. 262," accessed on 25 March 2019, www.darwinproject.ac.uk/DCP-LETT-262.

64 **"In the event":** King (1839), p. 23.

65 **A prodigious collector:** Desmond and Moore (1991), p. 129.

65 **On a stroll at dusk:** Darwin (2001), p. 402.

66 **"the mystery of mysteries":** Letter, John Herschel to Charles Lyell, February 20, 1836, quoted in W. H. Cannon (1961).

67 **"the most memorable event":** Darwin (2001), p. 427.

67 **"If there is the slightest":** Barlow (1963), p. 262.

67 **Darwin was most:** Sulloway (1982), p. 192, p. 359.

67 **"I never dreamed":** Darwin (1845), p. 394.

70 **"It at once struck":** Darwin (1887), p. 83.

70 **Darwin's reasons:** Van Whye (2007).

70 **"I determined":** Letter, Charles Darwin to Joseph Dalton Hooker, January 11, 1844, Darwin Correspondence Project, "Letter no. 729," accessed on 3 April 2019, www.darwinproject.ac.uk/DCP-LETT-729.

70 **"I am hard at work":** Letter, Charles Darwin to William Darwin Fox, March 19, 1855, Darwin Correspondence Project, "Letter no. 1651," accessed on 3 April 2019, www.darwinproject.ac.uk/DCP-LETT-1651.

72 **His own flock:** Secord (1981), p. 166.

72 **"I will show":** Letter, Charles Darwin to Charles Lyell, November 4, 1855, Darwin Correspondence Project, "Letter no. 1772," accessed on 3 April 2019, www.darwinproject.ac.uk/DCP-LETT-1772.

73 **The breeders did:** Secord (1981), p. 174.

73 **The two results:** Darwin (1859), p. 25–26.

73 **"I . . . never happened":** Darwin (1887), p. 87.

73 **"May not those":** Darwin (1859), p. 29.

74 **"I had no intention":** Letter, Charles Darwin to Asa Gray, May 22, 1860, Darwin Correspondence Project, "Letter no. 2814," accessed on 30 April 2019, www.darwinproject.ac.uk/DCP-LETT-2814.

75 **"paramount power":** Darwin (1868), p. 248.

75 **"the endless diversity":** Darwin (1862), p. 349.

76 **"I have just received":** Letter, Charles Darwin to Joseph Dalton Hooker, January 25, 1862, Darwin Correspondence Project, "Letter no. 3411," accessed on 30 April 2019, www.darwinproject.ac.uk/DCP -LETT-3411.

76 **Darwin discovered:** Arditti et al. (2012).

76 **"What a proboscis":** Letter, Charles Darwin to Joseph Dalton Hooker, January 30, 1862, Darwin Correspondence Project, "Letter no. 3421," accessed on 30 April 2019, www.darwinproject.ac.uk/DCP-LETT-3421.

76 **"In Madagascar there":** Darwin (1862), p. 188.

76 **The orchid and:** Darwin (1862), p. 202–203; Arditti et al. (2012), p. 425–427.

77 **Forty-one years:** Arditti et al. (2012).

77 **More than a century:** Alexandersson and Johnson (2002).

78 **The sword-billed hummingbird:** cited in Muchhala and Thomson (2009).

78 **the nectar bat:** Muchhala (2006).

78 **Darwin plainly admitted:** Darwin (1859), p. 131, p. 198.

78 **However, at many points**: See C. Johnson (2015) and J. Beatty (2006) for broad discussions of Darwin's insight into the nature of variations.

78 **"[A]s variations manifestly":** Darwin (1859), p. 41.

78 **"forms existing in larger":** Darwin (1859), p. 177.

79 **"I can see no reason":** Darwin (1859), p. 94.

79 **"[W]ith animals":** Darwin (1859), p. 242.

79 **"Asa Gray and some":** Letter, Charles Darwin to Frances Julia (Snow) Wedgwood, July 11, 1861, Darwin Correspondence Project, "Letter no. 3206," accessed on 7 April 2019, www.darwinproject.ac.uk/DCP-LETT-3206.

CHAPTER 4

PAGE

82 **On the other hand:** "The Queen's Visit to the Britannia Tubular Bridge." *The Victorian Web.* 7 Oct 2017. www.victorianweb.org/history/victoria /16.html

83 **The Post had:** Rabbe, Will. "The Washington Post's Famous 1915 Typo." *MSNBC.* 6 Aug 2013. www.msnbc.com/hardball/the-washington-posts -famous-1915-typo

83 **The blasphemy was:** Brown, DeNeen L. "The Bible Museum's 'Wicked Bible': Thou Shalt Commit Adultery." *Washington Post.* 18 Nov 2017. www. washingtonpost.com/news/retropolis/wp/2017/11/17/the-new-bible

-museums-wicked-bible-thou-shalt-commit-adultery/?utm_term=.
f1e9f851f41f:
In this case, the pair of typos are suspected to not be random errors. The blasphemy might have been a matter of sabotage by a rival printer.

84 **The original text:** Wain et al. (2007).

85 **It was difficult:** Lederberg and Lederberg (1952).

88 **Watson and Crick:** Watson and Crick (1953a).

90 **Watson took a celebratory:** Carroll (2013), p. 332.

91 *E. coli* **has a single:** Blattner et al. (1997).

91 **To put that:** "Average Typing Speed Infographic." *Ratatype.* www.ratatype.com/learn/average-typing-speed/

91 **Detailed studies:** Fijalkowska et al. (2012).

92 **The range is about:** Roach et al. (2010); Conrad et al. (2011).

92 **Stephen Quake:** Wang et al. (2012).

92 **Such studies have:** see for example Zhu et al. (2014). Note the frequency of all possible base substitutions are not equal; changing bases of similar size and shape (e.g. A to G, C to T) are more frequent than changes in bases of different size (e.g. A to C, G to T).

95 **"spontaneous mutation may":** Watson and Crick (1953b), p. 966.

95 **They imagined that:** Watson and Crick (1953c). Note, Monod was aware of this proposal (Monod, 1971, p. 192) but no empirical evidence existed to support it.

95 **But biochemists have:** Bebenek et al. (2011); Wang et al. (2011); Kimsey et al. (2015); Kimsey et al. (2018).

CHAPTER 5

PAGE

97 **"Name the greatest":** Twain (1935), p. 374.

97 **The law was upheld:** "Tennessee Evolution Statutes." Chapter 27, House Bill No. 185 (1925); Chapter 237, House Bill No. 48 (1967). *Famous Trials in American History: Tennessee vs. John Scopes The "Monkey Trial."* law2.umkc.edu/faculty/Projects/FTrials/scopes/TENNSTAT.HTM.

98 **The process had:** Rossiianov (2002).

98 **"exclusively important problem":** Rossiianov (2002), p. 286.

98 **"to support the doctrine":** "Soviet Backs Plan to Test Evolution." *New York Times.* 17 June 1926.

99 **"exceptional scientific":** Rossiianov (2002), p. 302.

99 **The Ivanov episode:** Etkind (2008).

100 **You may be surprised:** Nei (2013).

100 **"I can see no limit":** Darwin (1859), p. 109.

100 **"chance alone is":** Monod (1971), p. 112.

106 **The pigeon genome:** Shapiro et al. (2013).

106 **Nevertheless, biologist Mike:** Shapiro et al. (2013).

108 **Interestingly, Shapiro:** Vikrey et al. (2015).

109 **The superb fossil record:** Campbell et al. (2010).

110 **Perhaps the most famous:** Hawkes et al. (2011).

110 **Similarly, the Andean:** Jessen et al. (1991).

110 **Examinations of the hemoglobin:** Storz (2016).

111 **In these birds:** see Zhu et al. (2018).

112 **A key clue came:** Baardsnes and Davies (2001).

114 **That detective work:** Deng et al. (2010).

115 About 23 percent of all possible new mutations do not alter the encoded amino acid, but the three-quarters figure I have given is higher because most mutations that accumulate and are not lost/purged from populations do not change protein sequences.

116 **For example, if a mutation:** Nei (2013).

118 **For example, one:** Huang et al. (2019).

119 **That means that:** Varki and Altheide (2005).

120 **They have uncovered:** Hedges et al. (2015); Marin et al. (2018).

120 **"The lineage splitting":** Hedges et al. (2015).

120 **protein KKKYMMKHL that changed:** Wain et al. (2007).

121 **The one change (M→R):** Wain et al. (2007).

121 **HIV-2, the lesser-known:** Sharp and Hahn (2011).

121 **the origin of the 2002–2004 severe acute respiratory syndrome:** Song et al. (2005).

121 **and of multiple outbreaks of Middle East Respiratory Syndrome:** Mohd, Tawfiq, and Memish (2016).

122 **Indeed, many of humanity's:** Babkin and Babkina (2015).

122 **and measles:** Furuse, Suzuki, and Oshitani (2010).

122 **The SARS-CoV-2 coronavirus:** Andersen et al. (2020); Zhang et al. (2020).

CHAPTER 6

PAGE

127 **But when Jamie Coots:** Tribell, William S. "Last Film Footage of Pastor Jamie Coots." 23 Mar 2014. Video. www.youtube.com /watch?v=5f4NyYqHXa8

128 **"To me that's what":** Wilking, Spencer and Lauren Effron. "Snake-Handling Pentecostal Pastor Dies From Snake Bite." *ABCNews.* 17 Feb 2014. abcnews.go.com/US/snake-handling-pentecostal-pastor-dies-snake -bite/story?id=22551754

129 **"It was the most":** Chang, Juju and Spencer Wilking. "Pentecostal Pastors Argue 'Snake Handling' Is Their Religious Right." *ABCNews.* 21 Nov 2013. abcnews.go.com/US/pentecostal-pastors-argue-snake-handling -religious/story?id=20971576

129 **"It is an inner peace":** Chang, Juju and Spencer Wilking. "Pentecostal Pastors Argue 'Snake Handling' Is Their Religious Right." *ABCNews.* 21 Nov. 2013 abcnews.go.com/US/pentecostal-pastors-argue-snake-handling -religious/story?id=20971576.

129 **Only three months:** Wilking, Spencer and Lauren Effron. "Snake-Handling Pentecostal Pastor Dies From Snake Bite." *ABCNews.* 17 Feb

2014. abcnews.go.com/US/snake-handling-pentecostal-pastor-dies
-snake-bite/story?id=22551754

129 **second of three Pentecostal snake-handlers:** Duin, Julia. "Serpent
-Handling Pastor Profiled Earlier in Washington Post Dies from Rattlesnake Bite." *Washington Post.* 29 May 2012. www.washingtonpost.com
/lifestyle/style/serpent-handling-pastor-profiled-earlier-in-washington
-post-dies-from-rattlesnake-bite/2012/05/29/gJQAJef5zU_story.html
?utm_term=.a321063ededb; "Kentucky Man Dies from Snake Bite at
Church Service." *CBS News.* 28 July 2015. cbsnews.com/news/kentucky
-man-dies-from-snake-bite-at-church-service/

129 **Reviews of hospital records:** Pauly, Greg. "Misplaced Fears: Rattlesnakes Are Not as Dangerous as Ladders, Trees, Dogs, or Large
TVs." *Natural History Museum of Los Angeles County.* nhm.org/stories
/misplaced-fears-rattlesnakes-are-not-dangerous-ladders-trees-dogs
-or-large-tvs

129 **Now contrast this:** "Facts + Statistics: Mortality Risk." *Insurance Information Institute.* www.iii.org/fact-statistic/facts-statistics-mortality-risk

134 **Of these about twenty:** Walter et al. (1998).

135 **A recent study of Icelanders:** Jónsson et al. (2017).

135 **For example, at least thirty:** Iossifov et al. (2014).

135 **A study of over 5.7 million:** Sandin et al. (2016).

135 **Largely because of trisomy:** Hardy and Hardy (2015).

136 **He eventually returned to school:** Green, Jesse. "The Man Who Was
Immune to AIDS." 5 June 2014. *New York Magazine.* nymag.com/health
/bestdoctors/2014/steve-crohn-aids-2014–6/

136 **Later that summer:** "Thirty Years of HIV/AIDS: Snapshots of an Epidemic." *amfAR: The Foundation for AIDS Research.* amfar.org/thirty-years
-of-hiv/aids-snapshots-of-an-epidemic/

137 **But as yet still:** Green, Jesse. "The Man Who Was Immune to AIDS." 5
June 2014. *New York Magazine.* nymag.com/health/bestdoctors/2014
/steve-crohn-aids-2014–6/

137 **"I could not infect his CD4":** Pincock (2013).

137 **Sequencing of Crohn's:** Liu et al. (1996).

138 **A decade later:** Van Der Ryst (2015).

138 **That same year:** Brown (2015).

138 **Extensive global surveys:** Solloch et al. (2017).

138 **The delta32 mutation:** Hummel et al. (2005).

146 **It estimated that altogether:** Glanville et al. (2009).

146 **There, he was put:** Roberts, Joe. "Pastor almost killed by snake during
sermon vows to keep handling snakes." *MetroUK.* 24 Aug 2018. metro.
co.uk/2018/08/24/pastor-almost-killed-by-snake-during-sermon-vows
-to-keep-handling-snakes-7878534/; Barcroft TV. "Pastor Fights For Life
After Deadly Rattlesnake's Bite | MY LIFE INSIDE: THE SNAKE CHURCH."
23 Aug 2018. Video. www.youtube.com/watch?v=7ewdisyzk4k

146 **There, he was put:** Coots received antivenom according to Thomas
Midlane, Senior Producer. Email to author, August 2, 2019.

CHAPTER 7

PAGE

149 **"The fire was bouncing":** Burchard, Hank. "Lightning Strikes 4 Times." 2 May 1972. *The Ledger.* Vol 65, No 200. Lakeland, Florida. Page 3D.

150 **About ten per cent:** "Lightning Facts and Statistics." *Weather Imagery.* 18 Feb 2007. www.weatherimagery.com/blog/lightning-facts/

150 **The third strike:** Burchard, Hank. "Lightning Strikes 4 Times." 2 May 1972. *The Ledger.* Vol 65, No 200. Lakeland, Florida. Page 3D.

150 **"the human lightning conductor":** "Most Lighting Strikes Survived." *Guinness World Records.* www.guinnessworldrecords.com/world-records/most-lightning-strikes-survived

150 **"Lightning Man" or the "Spark Ranger":** Dunkel, Tom. "Lightning Strikes: A Man Hit Seven Times." *The Washington Post.* 15 Aug 2013. www.washingtonpost.com/lifestyle/magazine/inside-the-life-of-the-man-known-as-the-spark-ranger/2013/08/15/947cf2d8-ea40-11e2-8f22-de4bd2a2bd39_story.html

150 **Sullivan was zapped:** Dunkel, Tom. "Lightning Strikes: A Man Hit Seven Times." *The Washington Post.* 15 Aug 2013. www.washingtonpost.com/lifestyle/magazine/inside-the-life-of-the-man-known-as-the-spark-ranger/2013/08/15/947cf2d8-ea40-11e2-8f22-de4bd2a2bd39_story.html

151 **The curves reflect:** Nordling (1953).

151 **about two of every five:** "Cancer Statistics." *National Cancer Institute.* National Institutes of Health. www.cancer.gov/about-cancer/understanding/statistics

152 **But because cancer:** Nordling (1953); Armitage and Doll (1954).

154 **Over the ensuing decades:** Nowell (1976).

155 **Over several decades:** Vogelstein et al. (2013).

157 **And typically, more:** Vogelstein et al. (2013).

159 **At each end:** Govindan et al. (2012).

159 **Cigarette smoke contains:** DeMarini (2004).

159 **Analyzing normal-appearing:** Martincorena et al. (2015).

160 **Mutations occur at random:** I am again using random in the broad sense; certain kinds of mutations are more likely/frequent than others (see Chapter 4).

160 **It turns out these:** Vogelstein et al. (2013).

161 **For example, smoking increases:** Samet et al. (2009).

161 **"The public health authorities":** Vonnegut (1950), p. xv.

161 **Similarly, more people:** Green et al. (2011); Watts et al. (2018).

161 **"Use sunscreen!":** "Vonnegut. Agnes Scott College Commencement Address." 15 May 1999. *C-SPAN.* Transcript. www.c-span.org/video/?123554-1/agnes-scott-college-commencement-address; Vonnegut (1999), p. 12.

162 **Today, there are scores:** "Targeted Cancer Therapies." *National Cancer Institute.* National Institutes of Health. www.cancer.gov/about-cancer/treatment/types/targeted-therapies/targeted-therapies-fact-sheet.

AFTERWORD

PAGE

163 **"Tiger got to hunt":** Vonnegut, *Cat's Cradle*, (2010), p. 182.

163 **"(H)ers would have":** Vonnegut, *Slapstick*, (2010), p. 12–13.

164 **Two coaches of:** Maggie Bartel and Henry Lee. "New Jersey Train Plunges off a Bridge into Newark Bay Killing More than 40 in 1958." *Daily News.* 16 Sept 1958. *New York Daily News.* 15 Sept 2015. www.nydailynews.com/news/national/tragic-train-accident-new-jersey-1958-article-1.2360153

164 **"Since Alice had never":** Vonnegut, *Slapstick*, (2010), p. 14.

165 **But our chance–driven:** Jungersen (2013), p. 266.

165 **"Everyone now knows":** Vonnegut, *Sirens*, (2009), p. 1.

166 **"Puny man can":** Vonnegut, *Sirens*, (2009), p. 183.

166 **"I was a victim":** Vonnegut, *Sirens*, (2009), p. 232.

A CONVERSATION ABOUT CHANCE

Verbatim quotes in italics. Verbatim quotes directly to author in italics and underlined. Words created and inserted by author are in bold, plain text.

PAGE

168 *MacFARLANE Not especially:* Rabin, Nathan. "Seth MacFarlane." *AV Club.* 26 Jan 2005. *The Onion.* tv.avclub.com/seth-macfarlane-1798208419.

168 *SILVERMAN I am insanely lucky:* Sarah Silverman. "Hi. This is me telling everyone in my life. . . ." 6 July 2016. www.facebook.com/SarahSilverman/posts/hi-this-is-me-telling-everyone-in-my-life-at-once-why-i-havent-been-around-this-/1160269914023100/

168 *SILVERMAN I didn't know going in:* Murfett, Andrew. "After a Near-Death Experience, Comedy Alleviates Sarah Silverman's PTSD Pain." *The Sydney Morning Herald.* 25 May 2017. www.smh.com.au/entertainment/tv-and-radio/after-a-neardeath-experience-comedy-alleviates-sarah-silvermans-pain-20170525-gwd4c0.html

168 *SILVERMAN Don't even know why:* Sarah Silverman. "Hi. This is me telling everyone in my life . . ." 6 July 2016. www.facebook.com/SarahSilverman/posts/hi-this-is-me-telling-everyone-in-my-life-at-once-why-i-havent-been-around-this-/1160269914023100/

168 *SILVERMAN He looks down my throat:* Silverman, Sarah, writer, performer. *A Speck of Dust.* Directed by Liam Lynch. Netflix Studios, Eleven Eleven O'Clock Productions, Jash Network, 2017.

168 *SILVERMAN When I woke up 5 days later:* Sarah Silverman. "Hi. This is me telling everyone in my life . . ." 6 July 2016. www.facebook.com/SarahSilverman/posts/hi-this-is-me-telling-everyone-in-my-life-at-once-why-i-havent-been-around-this-/1160269914023100/

169 *SILVERMAN I spent the first two days home:* Sarah Silverman. "Hi. This is me telling everyone in my life . . ." 6 July 2016. www.facebook.com

/SarahSilverman/posts/hi-this-is-me-telling-everyone-in-my-life-at -once-why-i-havent-been-around-this-/1160269914023100/

169 **SILVERMAN All of this has maybe made me:** Murfett, Andrew. "After a Near-Death Experience, Comedy Alleviates Sarah Silverman's PTSD Pain." *The Sydney Morning Herald.* 25 May 2017. www.smh.com.au/ entertainment/tv-and-radio/after-a-neardeath-experience-comedy -alleviates-sarah-silvermans-pain-20170525-gwd4c0.html

169 **MONOD I, too, am grateful:** Based on Carroll (2013), p. 36–37.

169 **CAMUS Well, yes. It was quite dangerous.**

169 **CAMUS I did not get caught:** Based on Carroll (2013), p. 220.

170 **MONOD The Gestapo was everywhere:** Based on interview with Olivier Monod, Paris, August 17, 2010; Carroll (2013), p. 208–209.

170 **VONNEGUT I got caught.**

170 **VONNEGUT We were in this gully:** Hayman et al. (1977).

170 **VONNEGUT So we did.**

170 **VONNEGUT They said the war was all over:** Hayman et al. (1977).

170 **VONNEGUT Yes and no. Six weeks later . . .**

170 **VONNEGUT We never expected to get it:** Hayman et al. (1977).

171 **VONNEGUT Twenty-five years later I wrote Slaughterhouse-Five.**

171 **IDLE He served in the Royal Air Force:** Based on Idle (2018), p. 4.

171 **GERVAIS So sorry, mates:** Based on Adams, Tim. "Second Coming." *The Guardian.* 10 July 2005. www.theguardian.com/theobserver/2005 /jul/10/comedy.television

172 **GERVAIS But it was a long time . . .**

172 **GERVAIS I remember saying to my mum:** Gordon, Bryony. "Ricky Gervais: Don't Ask Me the Price of Milk—I Fly by Private Jet." *The Telegraph.* 26 Aug 2011. www.telegraph.co.uk/culture/comedy/8722858/Ricky -Gervais-Dont-ask-me-the-price-of-milk-I-fly-by-private-jet.html.

172 **IZZARD World War II is it:** "Eddie Izzard Explains Why He Doesn't Believe in Religion." *Skaylan.* 24 Nov 2018. *YouTube.* www.youtube.com /watch?v=emcb_HqBLrU

172 **IDLE It saves time:** Eric Idle. Email reply to author, 8/21/19.

172 **MAHER We are talking about humanistic gods:** Gettelman, Elizabeth. "The MoJo Interview: Bill Maher." *Mother Jones.* September/October 2008. www.motherjones.com/media/2008/09/mojo-interview-bill-maher/

173 **IZZARD: Religion is just full of stories:** Yaqhubi, Zohra D. "Izzard Accepts Humanist Award." *The Harvard Crimson.* 21 Feb 2013. www.thecrimson .com/article/2013/2/21/izzard-accepts-humanist-award/

173 **MAHER We are used to the story:** Gettelman, Elizabeth. "The MoJo Interview: Bill Maher." *Mother Jones.* September/October 2008.

173 **IDLE I think we [Monty Python] should now apply:** Silman, Anna. "Monty Python Reunites at Tribeca: "We are Funnier than Scientology . . . We Could be a Religion." *Salon.* 25 April 2015. www.salon .com/2015/04/24/monty_python_reunites_at_tribeca_we_are_funnier _than_scientology_we_could_be_a_religion

173 **IDLE Comedy is telling the truth:** Eric Idle. Email reply to author, 8/21/19.

173 **MAHER When people laugh:** Gettelman, Elizabeth. "The MoJo Interview: Bill Maher." *Mother Jones.* September/October 2008.

173 **VONNEGUT I agree.**

173 **VONNEGUT The telling of jokes:** Rentilly, J. "The Best Jokes Are Dangerous, An Interview with Kurt Vonnegut, Part Three." *McSweeney's.* 16 Sept 2002. www.mcsweeneys.net/articles/the-best-jokes-are-dangerous-an-interview-with-kurt-vonnegut-part-three

174 **IDLE Truth is the point of comedy:** Eric Idle. Email reply to author, 8/21/19.

174 **SILVERMAN I don't know if there is a God:** Silverman, Sarah, writer, performer. *A Speck of Dust.* Directed by Liam Lynch. Netflix Studios, Eleven Eleven O'Clock Productions, Jash Network, 2017.

174 **IZZARD Religious people might think:** Adams, Tim. "Eddie Izzard: 'Everything I Do in Life is Trying to Get My Mother Back.'" *The Guardian.* 10 Sept 2017. www.theguardian.com/culture/2017/sep/10/eddie-izzard-trying-to-get-mother-back-victoria-and-abdul

174 **GERVAIS Since there is nothing:** Gervais, Ricky. "Does God Exist? Ricky Gervais Takes Your Questions." *The Wall Street Journal.* 22 Dec 2010. blogs.wsj.com/speakeasy/2010/12/22/does-god-exist-ricky-gervais-takes-your-questions/

175 **MacFARLANE the traditional sci-fi method of storytelling:** "Seth MacFarlane on Using Science Fiction to Explore Humanity." *Sean Carroll's Mindscape.* 5 Aug 2019. www.preposterousuniverse.com/podcast/2019/08/05/58-seth-macfarlane-on-using-science-fiction-to-explore-humanity/

175 **VONNEGUT When Shakespeare figured:** Standish, David. "Kurt Vonnegut: The Playboy Interview." *Playboy.* 20 July 1973.

175 **CAMUS There is but one truly serious philosophical problem:** Camus (1991), p. 3.

175 **VONNEGUT So there's another barrel:** Vonnegut (2014), p. 64.

175 **IDLE Because Philosophy like wine is better in French:** Eric Idle. Email reply to author, 8/21/19.

176 **IDLE It won the Grand Jury Prize in 1983:** Eric Idle. Email reply to author, 8/21/19.

176 **MAHER In a culture that needs caffeine-free:** Gettelman, Elizabeth. "The MoJo Interview: Bill Maher." *Mother Jones.* September/October 2008. www.motherjones.com/media/2008/09/mojo-interview-bill-maher/

176 **IDLE Well the point about science:** Eric Idle. Email reply to author, 8/21/19.

176 **MacFARLANE It's like the civil-rights movement:** Woods, Stacey Grenrock. "Hungover with Seth MacFarlane." *Esquire.* 18 Aug 2009. www.esquire.com/entertainment/movies/a6099/seth-macfarlane-interview-0909/

176 **VONNEGUT I love science:** Vonnegut (2014), p. 59.

176 **VONNEGUT We once had a memorial service:** Vonnegut (2014), p. 57.

176 **GERVAIS Science seeks the truth:** Gervais, Ricky. "Ricky Gervais: Why I'm an Atheist." *The Wall Street Journal.* 19 Dec 2010. blogs.wsj.com

/speakeasy/2010/12/19/a-holiday-message-from-ricky-gervais-why
-im-an-atheist/

177 *IZZARD There is a rule in every major religion:* Izzard (2018), p. 335.

177 *GERVAIS "Do unto others . . .":* Gervais, Ricky. "Ricky Gervais: Why
I'm an Atheist." *The Wall Street Journal.* 19 Dec 2010. blogs.wsj.com
/speakeasy/2010/12/19/a-holiday-message-from-ricky-gervais-why-im
-an-atheist/

177 *MAHER No ethicist has a problem:* Gettelman, Elizabeth. "The MoJo
Interview: Bill Maher." *Mother Jones.* September/October 2008. www
.motherjones.com/media/2008/09/mojo-interview-bill-maher/

177 *GERVAIS And that's where spirituality:* Gervais, Ricky. "Ricky Gervais:
Why I'm an Atheist." *The Wall Street Journal.* 19 Dec 2010. blogs.wsj.com/
speakeasy/2010/12/19/a-holiday-message-from-ricky-gervais-why-im
-an-atheist/

178 *VONNEGUT We are here on Earth:* Vonnegut (2007), p. 62.

178 **CAMUS Maybe so. But of all the things:** Based on Camus (1991).

178 *GERVAIS You should bring something:* Turner, N.A. "Ricky Gervais on
Chasing Your Dream, Doing the Work and Living a Creative Life." *Medium.*
4 Aug 2019. medium.com/swlh/ricky-gervais-on-chasing-your-dream
-doing-the-work-and-living-a-creative-life-946684bd634d

178 **MONOD We must create both things and ideas:** Based on BBC Interview with Gerald Leach in Carroll (2013) p. 482; Monod (1971), p. xi.;
Monod (1969).

178 *GERVAIS It always comes back to us:* Wooley, Charles. "When Charles
Wooley met Ricky Gervais." *60 Minutes.* 2019. 9now.nine.com.au/60
-minutes/charles-wooley-ricky-gervais-seriously-funny/1b49b447
-bdd2-43c2-8238-b4c2ba287646

BIBLIOGRAPHY AND FURTHER READING

FURTHER READING

The roles of chance and contingency in nature and in human affairs have been examined in several works over the past half-century. For those interested in exploring these topics further, I recommend the following outstanding and accessible books:

Alvarez, Walter. (2017) *A Most Improbable Journey: A Big History of Our Planet and Ourselves.* New York: W.W. Norton & Company.

Conway-Morris, Simon. (2003) *Life's Solution: Inevitable Humans in a Lonely Universe.* Cambridge: Cambridge University Press.

Dawkins, Richard. (1986) *The Blind Watchmaker: Why the Evidence of Evolution Reveals a Universe Without Design.* New York: W.W. Norton & Company.

Gould, Stephen Jay. (1989) *Wonderful Life: The Burgess Shale and the Nature of History.* New York: W.W. Norton & Company.

Losos, Jonathan. (2018) *Improbable Destinies: Fate, Chance, and the Future of Evolution.* New York: Riverhead Books.

Monod, Jacques. (1971) *Chance and Necessity: An Essay on the Natural Philosophy of Modern Biology.* New York: Alfred A. Knopf.

Taleb, Nassim Nicholas. (2004) *Fooled by Randomness: The Hidden Role of Chance in Life and the Markets.* New York: Random House.

BIBLIOGRAPHY

ACIA (2004). *Impacts of a Warming Arctic: Arctic Climate Impact Assessment. ACIA Overview report.* Cambridge University Press.

Alexandersson, R. and S.D. Johnson. (2002) "Pollinator-Mediated Selection on Flower-Tube Length in a Hawkmoth-Pollinated Gladiolus (Iridaceae)." *Proceedings of the Royal Society B: Biological Sciences.* 269(1491): 631–636.

Alley, Richard B. (2000) "The Younger Dryas Cold Interval as Viewed from Central Greenland." *Quaternary Science Reviews.* 19: 213–226.

Alley, R.B., D.A. Meese, C.A. Shuman, et al. (1993) "Abrupt Increase in Greenland Snow Accumulation at the End of the Younger Dryas Event." *Nature*. 362: 527–529.

Alvarez, L., et al. (1980) "Extraterrestrial Cause for the Cretaceous-Tertiary Extinction: Experimental results and theoretical interpretation." *Science*. 208: 1095–1108.

Alvarez, Walter. (1997) *T. rex and the Crater of Doom*. Princeton, New Jersey: Princeton University Press.

Anagnostou, Eleni, Eleanor H. John, Kirsty M. Edgar, et al. (2016) "Changing Atmospheric CO2 Concentration was the Primary Driver of Early Cenozoic Climate." *Nature*. 533: 380–384.

Andersen, Kristian G., Andrew Rambaut, W. Ian Lipkin, et al. (2020) "The Proximal Origin of SARS-CoV-2." *Nature Medicine*. Published online. https://doi.org/10.1038/s41591-020-0820-9.

Arditti, Joseph, John Elliott, Ian J. Kitching, and Lutz T. Wasserthal. (2012) "'Good Heavens What Insect Can Suck It'—Charles Darwin, *Angraecum sesquipedale* and *Xanthopan morganii praedicta*." *Botanical Journal of the Linnean Society*. 169: 403–432.

Armitage, P. and R. Doll. (1954) "The Age Distribution of Cancer and a Multi-Stage Theory of Carcinogenesis." *British Journal of Cancer*. 8(1): 1–12.

Artemieva, Natalia, Joanna Morgan, and Expedition 364 Science Party. (2017) "Quantifying the Release of Climate-Active Gases by Large Meteorite Impacts With a Case Study of Chicxulub." *Geophysical Research Letters*. 44(20): 10180–10188.

Baardsnes, Jason and Peter L. Davies. (2001) "Sialic Acid Synthase: The Origin of Fish Type III Antifreeze Protein?" *Trends in Biochemical Sciences*. 26(8): 468–469.

Babkin, Igor V. and Irina N. Babkina. (2015) "The Origin of the Variola Virus." *Viruses*. 7: 1100–1112.

Barker, Stephen, Gregor Knorr, R. Lawrence Edwards, et al. (2011) "800,000 Years of Abrupt Climate Variability." *Science*. 334: 347–351.

Barlow, Emma Nora. (1963) "Darwin's Ornithological Notes." *Bulletin of the British Museum (Natural History) Historical Series*. 2: 201–273. Accessed: darwin-online.org.uk/content/frameset?itemID=F1577&viewtype=text&pageseq=1

Bax, Ben, Chun-wa Chung, and Colin Edge. (2017) "Getting the Chemistry Right: Protonation, Tautomers and the Importance of H Atoms in Biological Chemistry." *Acta Crystallographica Section D Structural Biology*. 73(Pt 2): 131–140.

Beatty, John. (2006) "Chance Variation: Darwin on Orchids." *Philosophy of Science*. 73(5): 629–641.

Beatson, K., M. Khorsandi, and N. Grubb. (2013) "Wolff–Parkinson–White Syndrome and Myocardial Infarction in Ventricular Fibrillation Arrest: A Case of Two One-Eyed Tigers." *QJM: An International Journal of Medicine*. 106(8): 755–757.

Bebenek, Katarzyna, Lars C. Pedersen, and Thomas A. Kunkel. (2011) "Replication Infidelity via a Mismatch with Watson-Crick Geometry." *Proceedings of the National Academy of Sciences*. 108(5): 1862–1867.

Behrensmeyer, Anna K., Richard Potts, and Alan Deino. (2018) "The Oltulelei Formation of the Southern Kenyan Rift Valley: A Chronicle of Rapid Landscape Transformation over the Last 500 k.y." *Geological Society of America Bulletin*. 130: 1474–1492.

Blattner, F.R., G. Plunkett 3rd, C.A. Bloch, et al. (1997) "The Complete Genome Sequence of *Escherichia coli* K-12." *Science*. 277(5331): 1453–1462.

Blois, Jessica L. and Elizabeth A. Hadly. (2009) "Mammalian Response to Cenozoic Climatic Change." *Annual Review of Earth and Planetary Sciences*. 37: 8.1–8.28.

Bolhuis, Johan, Ian Tattersall, Noam Chomsky, et al. (2014) "How Could Language Have Evolved." *PLoS Biology*. 12(8): e1001934.

Bouilhol, Pierre, Oliver Jagoutz, John M. Hanchar, and Francis O. Dudas. (2013) "Dating the India-Eurasia Collision Through Arc Magmatic Records." *Earth and Planetary Science Letters*. 366: 163–175.

Broecker, W.S. (1995) "Cooling the Tropics." *Nature*. 376: 212–213.

Broecker, W.S. (2015) "The Collision That Changed the World." *Elementa: Science of the Anthropocene*. 3. 000061. 10.12952/journal.elementa.000061.

Brooks, Alison S., John E. Yellen, Richard Potts, et al. (2018) "Long-Distance Stone Transport and Pigment Use in the Earliest Middle Stone Age." *Science*. 360: 90–94.

Brown, Timothy Ray. (2015) "I Am the Berlin Patient: A Personal Reflection." *AIDS Research and Human Retroviruses*. 31(1): 2–3.

Brugger, Julia, Georg Feulner, and Stefan Petri. (2017) "Baby, It's Cold Outside: Climate Model Simulations of the Effects of the Asteroid Impact at the End of the Cretaceous." *Geophysical Research Letters*. 44: 419–427.

Brusatte, Stephen L., Jingmai K. O'Connor, and Erich D. Jarvis. (2015) "The Origin and Diversification of Birds." *Current Biology*. 25(19): R888-R898.

Campbell, Kevin L, Jason E. E. Roberts, Laura N. Watson, et al. (2010) "Substitutions in Woolly Mammoth Hemoglobin Confer Biochemical Properties Adaptive for Cold Tolerance." *Nature Genetics*. 42(6): 536–540.

Camus, Albert. (1991) *The Myth of Sisyphus and Other Essays*. Translated by Justin O'Brien, *Le Mythe de Sisyphe*, 1942. First Vintage International Edition, 1991. New York: Random House.

Cannon, Walter F. (1961) "The Impact of Uniformitarianism: Two Letters from John Herschel to Charles Lyell, 1836–1837." *Proceedings of the American Philosophical Society*. 105(3): 301–314.

Carroll, Sean B. (2013) *Brave Genius: A Scientist, A Philosopher, and Their Daring Adventures from the French Resistance to the Nobel Prize*. New York: Crown Publishers.

Carter, Andrew, Teal R. Riley, Claus-Dieter Hillenbrand, and Martin Rittner. (2017) "Widespread Antarctic Glaciation During the Late Eocene." *Earth and Planetary Science Letters*. 458: 49–57.

Clark, Luke, Andrew J. Lawrence, Frances Astley-Jones, and Nicola Gray. (2009) "Gambling Near-Misses Enhance Motivation to Gamble and Recruit Win-Related Brain Circuitry." *Neuron.* 61: 481–490.

Conrad, Donald F., Jonathan E.M. Keebler, Mark A. DePristo, et al. (2011) "Variation in Genome-Wide Mutation Rates Within and Between Human Families." *Nature Genetics.* 43(7): 712–714.

Croson, Rachel and James Sundali. (2005) "The Gambler's Fallacy and the Hot Hand: Empirical Data from Casinos." *The Journal of Risk and Uncertainty.* 30(3): 195–209.

Cuppy, Will. (1941) *How to Become Extinct.* New York: Dover Publications.

Darwin, Charles R. (1845) *Journal of Researches into the Natural History and Geology of the Countries Visited During the Voyage of H.M.S. Beagle Round the World.* London: Murray. 2d ed. Accessed: darwin-online.org.uk/content/frameset?itemID=F14&viewtype=text&pageseq=1

Darwin, Charles R. (1859) *On the Origin of Species by Means of Natural Selection, or the Preservation of Favoured Races in the Struggle for Life.* London: Murray. 1st ed. Accessed: darwin-online.org.uk/content/frameset?itemID=F373&viewtype=text&pageseq=1

Darwin, Charles R. (1862) *On the Various Contrivances by Which British and Foreign Orchids are Fertilised by Insects.* London: John Murray.

Darwin, Charles R. (1868) *Variation of Animals and Plants Under Domestication.* New York: E.L. Godkin & Co. 1st ed., Volume 2. Accessed: darwin-online.org.uk/content/frameset?itemID=F877.2&viewtype=text&pageseq=1

Darwin, Charles R. (1887) *The Life and Letters of Charles Darwin: Including an Autobiographical Chapter.* Volume 1. Ed. Francis Darwin. London: John Murray.

Darwin, Charles R. (2001) *Charles Darwin's Beagle Diary.* Ed. Keynes. Cambridge: Cambridge University Press. Accessed: darwin-online.org.uk/content/frameset?viewtype=text&itemID=F1925&pageseq=1

DeMarini, David M. (2004) "Genotoxicity of Tobacco Smoke and Tobacco Smoke Condensate: A Review." *Mutation Research.* 567: 447–474.

deMenocal, Peter B. (1995) "Plio-Pleistocene African Climate." *Science.* 270: 53–59.

Deng, Cheng, C.H. Christina Cheng, Hua Ye, et al. (2010) "Evolution of an Antifreeze Protein by Neofunctionalization Under Escape from Adaptive Conflict." *Proceedings of the National Academy of Sciences.* 107(50): 21593–21598.

Desmond, Adrian and James Moore. (1991) *Darwin: The Life of a Tormented Evolutionist.* New York: W.W. Norton & Co.

Ditlevsen, P.D., Katrine Krogh Andersen, A. Svensson. (2007) "The DO-Climate Events Are Probably Noise Induced: Statistical Investigation of the Claimed 1470 Years Cycle." *Climate of the Past.* 3(1): 129–134.

Etkind, Alexander. (2008) "Beyond Eugenics: The Forgotten Scandal of Hybridizing Humans and Apes." *Studies in History and Philosophy of Biological and Biomedical Sciences.* 39: 205–210.

Feng, Yan-Jie, David C. Blackburn, Dan Liang, et al. (2017) "Phylogenomics Reveals Rapid, Simultaneous Diversification of Three Major Clades of

Gondwanan Frogs at the Cretaceous-Paleogene Boundary." *Proceedings of the National Academy of Sciences*. E5864–E5870.

Fiedel, S.J. (2011) "The Mysterious Onset of the Younger Dryas." *Quaternary International*. 242: 262–266.

Field, Daniel J., Antoine Bercovici, Jacob S. Berv, et al. (2018) "Early Evolution of Modern Birds Structured by Global Forest Collapse at the End-Cretaceous Mass Extinction." *Current Biology*. 28: 1825–1831.

Fijalkowska, Iwona J., Roel M. Schaaper, and Piotr Jonczyk. (2012) "DNA Replication Fidelity in *Escherichia coli*: A Multi-DNA Polymerase Affair." *FEMS Microbiology Reviews*. 36: 1105–1121.

Filippelli, Gabriel M. and José-Abel Flores. (2009) "From the Warm Pliocene to the Cold Pleistocene: A Tale of Two Oceans." *The Geological Society of America*. 37(10): 959–960.

Furuse, Yuki, Akira Suzuki, and Hitoshi Oshitani. (2010) "Origin of Measles Virus: Divergence from Rinderpest Virus Between the 11th and 12th Centuries." *Virology Journal*. 7(52).

Gallala, Njoud, Dalila Zaghbib-Turki, Ignacio Arenillas, et al. (2009) "Catastrophic Mass Extinction and Assemblage Evolution in Planktic Foraminifera Across the Cretaceous/Paleogene (K/Pg) Boundary at Bidart (SW France)." *Marine Micropaleontology*. 72: 196–209.

Glanville, Jacob, Wenwu Zhai, Jan Berka, et al. (2009) "Precise Determination of the Diversity of a Combinatorial Antibody Library Gives Insight into the Human Immunoglobulin Repertoire." *Proceedings of the National Academy of Sciences*. 106(48): 20216–20221.

Goff, James, Catherine Chagué-Goff, Michael Archer, et al. (2012) "The Eltanin Asteroid Impact: Possible South Pacific Palaeomegatsunami Footprint and Potential Implications for the Pliocene-Pleistocene Transition." *Journal of Quaternary Science*. 27(7): 660–670.

Govindan, Ramaswamy, Li Ding, Malachi Griffith, et al. (2012) "Genomic Landscape of Non-Small Cell Lung Cancer in Smokers and Never-Smokers." *Cell*. 150: 1121–1134.

Gowlett, J.A.J. (2016) "The Discovery of Fire by Humans: A Long and Convoluted Process." *Philosophical Transactions of the Royal Society of London B*. 371: 20150164.

Grech, Victor. (2002) "Unexplained Differences in Sex Ratios at Birth in Europe and North America." *BMJ*. 324: 1010–1011.

Green, Adèle C., Gail M. Williams, Valerie Logan, and Geoffrey M. Strutton. (2011) "Reduced Melanoma After Regular Sunscreen Use: Randomized Trial Follow-Up." 29(3): 257–263.

Greenland Ice-Core Project (GRIP) Members. (1993) "Climate Instability During the Last Interglacial Period Recorded in the GRIP Ice Core." *Nature*. 364: 203–207.

Gulick, Sean P.S., Timothy J. Bralower, Jens Ormö, et al. (2019) "The First Day of the Cenozoic." *Proceedings of the National Academy of Sciences*. 116(39): 19342–19351.

Hansen, James E. and Makiko Sato. (2012) "Climate Sensitivity Estimated from Earth's Climate History." *NASA Goddard Institute for Space Studies and Columbia University Earth Institute*. 1–19.

Hansen, James, Makiko Sato, Gary Russell, and Pushker Kharecha. (2013) "Climate Sensitivity, Sea Level and Atmospheric Carbon Dioxide." *Philosophical Transactions of the Royal Society A*. 371: 20120294.

Hardy, Kathy and Philip John Hardy. (2015) "1st Trimester Miscarriage: Four Decades of Study." *Translational Pediatrics*. 4(2): 189–200.

Hawkes, Lucy A., Sivananinthaperumal Balachandran, Nyambayar Batbayar, et al. (2011) "The Trans-Himalayan Flights of Bar-Headed Geese (*Anser indicus*)." 108(23): 9516–9519.

Hayman, David, David Michaelis, George Plimpton, and Richard Rhodes. (1977) "Kurt Vonnegut, The Art of Fiction No. 64." *The Paris Review*. 69. www.theparisreview.org/interviews/3605/kurt-vonnegut-the -art-of-fiction-no-64-kurt-vonnegut

Hedges, S. Blair, Julie Marin, Michael Suleski, et al. (2015) "Tree of Life Reveals Clock-Like Speciation and Diversification." *Molecular Biology and Evolution*. 32(4): 835–845.

Henehan, Michael J., Andy Ridgwell, Ellen Thomas, et al. (2019) "Rapid Ocean Acidification and Protracted Earth System Recovery Followed the End-Cretaceous Chicxulub Impact." *Proceedings of the National Academy of the Sciences*. 201905989; DOI: 10.1073/pnas.1905989116.

Hildebrand, A.R. (1991) "Chicxulub Crater: A Possible Cretaceous/Tertiary Boundary Impact Crater on the Yucatán Peninsula, Mexico." *Geology*. 19: 867–871.

Hren, Michael T., Nathan D. Sheldon, Stephen T. Grimes, et al. (2013) "Terrestrial Cooling in Northern Europe During the Eocene-Oligocene Transition." *Proceedings of the National Academy of Sciences*. 110(19): 7562–7567.

Huang, Kai, Shijun Ge, Wei Yi, et al. (2019) "Interactions of Unstable Hemoglobin Rush with Thalassemia and Hemoglobin E Result in Thalassemia Intermedia." *Hematology*. 24(1): 459–466.

Hughes, David W. (2003) "The Approximate Ratios Between the Diameters of Terrestrial Impact Craters and the Causative Incident Asteroids." *Monthly Notices of the Royal Astronomical Society*. 338: 999–1003.

Huff, Darrell and Irving Geis. (1959) *How to Take a Chance*. New York: W.W. Norton.

Hummel, S., D. Schmidt, B. Kremeyer, et al. (2005) "Detection of the CCR5-Δ32 HIV Resistance Gene in Bronze Age Skeletons." *Genes and Immunity*. 6: 371–374.

Idle, Eric. (2018) *Always Look on the Bright Side of Life: A Sortabiography*. New York: Crown Archetype.

Iossifov, Ivan, Brian J. O'Roak, Stephan J. Sanders, et al. (2014) "The Contribution of De Novo Coding Mutations to Autism Spectrum Disorder." *Nature*. 515: 216–221.

Izzard, Eddie. (2018) *Believe Me: A Memoir of Love, Death and Jazz Chickens*. London: Michael Joseph.

Jarvis, Erich D., Siavash Mirarab, Andre J. Aberer, et al. (2014) "Whole-Genome Analyses Resolve Early Branches in the Tree of Life of Modern Birds." *Science*. 346 (6215): 1320–1331.

Jessen, Timm-H., Roy E. Weber, Giulio Fermi, et al. (1991) "Adaptation of Bird Hemoglobins to High Altitudes: Demonstration of Molecular Mechanism by Protein Engineering." *Proceedings of the National Academy of Sciences.* 88: 6519–6522.

Johnson, Curtis. (2015) *Darwin's Dice: The Idea of Chance in the Thought of Charles Darwin.* Oxford: Oxford University Press.

Jónsson, Hákon, Patrick Sulem, Birte Kehr, et al. (2017) "Parental Influence on Human Germline De Novo Mutations in 1,548 Trios from Iceland." *Nature.* 549: 519–522.

Jungersen, Christian. (2013) *You Disappear.* Translated from Danish by Misha Hoekstra. Anchor Books.

Kaiho, Kunio and Naga Oshima. (2017) "Site of Asteroid Impact Changed the History of Life on Earth: The Low Probability of Mass Extinction." *Scientific Reports.* 7: 14855.

Kennett, J.P. and L.D. Stott. (1991) "Abrupt Deep-Sea Warming, Palaeoceanographic Changes and Benthic Extinctions at the End of the Palaeocene." *Nature.* 353: 225–229.

Kimsey, Isaac J., Katja Petzold, Bharathwaj Sathyamoorthy, et al. (2015) "Visualizing Transient Watson-Crick-like Mispairs in DNA and RNA Duplexes." *Nature.* 519: 315–320.

Kimsey, Isaac J, Eric S. Szymanski, Walter J. Zahurancik, et al. (2018) "Dynamic Basis for dG • dT Misincorporation via Tautomerization and Ionization." 554: 195–201.

King, Philip Parker. (1839) *The Narrative of the Voyages of H.M. Ships Adventure and Beagle.* London: Colburn. 1st ed. 3 volumes & appendix: *Proceedings of the First Expedition, 1826–30.* Accessed: darwin-online.org.uk/content/frameset?itemID=F10.1&viewtype=text&pageseq=1

Koch, Paul L. and Anthony D. Barnosky. (2006) "Late Quaternary Extinctions: State of the Debate." *Annual Review of Ecology, Evolution, and Systematics.* 37: 215–250.

Kumar, Prakash, Xiaohui Yuan, M. Ravi Kumar, et al. (2007) "The Rapid Drift of the Indian Tectonic Plate." *Nature.* 449: 894–897.

Lacis, Andrew A, Gavin A. Schmidt, David Rind, and Reto A. Ruedy. (2010) "Atmospheric CO2: Principal Control Knob Governing Earth's Temperature." *Science.* 330: 356–359.

Leakey, L.S.B. (1974) *By the Evidence: Memoirs, 1932–1951.* New York: Harcourt Brace Janovich.

Lederberg, Joshua and Esther M. Lederberg. (1952) "Replica Plating and Indirect Selection of Bacterial Mutants." *Journal of Bacteriology.* 63(3): 399–406.

Lewis, John. (1974) *Beyond Chance and Necessity: A Critical Inquiry into Professor Jacques Monod's Chance and Necessity.* London: The Teilhard Centre for the Future of Man.

Liu, Rong, William A. Paxton, Sunny Shoe, et al. (1996) "Homozygous Defect in HIV-1 Coreceptor Accounts for Resistance of Some Multiply-Exposed Individuals to HIV-1 Infection." *Cell.* 86: 367–377.

Lohmann, Johannes and Peter D. Ditlevsen. (2018) "Random and Externally Controlled Occurrences of Dansgaard–Oeschger Events." *Climate of the Past.* 14: 609–617.

Longrich, N.R., J. Scriberas, and M.A. Wills. (2016) "Severe Extinction and Rapid Recovery of Mammals Across the Cretaceous–Palaeogene Boundary, and the Effects of Rarity on Patterns of Extinction and Recovery." *Journal of Evolutionary Biology.* 29: 1495–1512.

Lydekker, Richard. (1904) *Library of Natural History, Vol III.* Saalfield Publishing Company: Akron, OH.

Lyson, T.R., I.M. Miller, A.D. Bercovici, et al. (2019) "Exceptional Continental Record of Biotic Recovery After the Cretaceous-Paleogene Mass Extinction." *Science.* 366(6468): 977–983.

Marin, Julie, Giovanni Rapacciuolo, Gabriel C. Costa, et al. (2018) "Evolutionary Time Drives Global Tetrapod Diversity." *Proceedings of the Royal Society B.* 285: 20172378.

Martincorena, Iñigo, Amit Roshan, Moritz Gerstung, et al. (2015) "High Burden and Pervasive Positive Selection of Somatic Mutations in Normal Human Skin." *Science.* 348(6237): 880–886.

Martínez-Botí, M.A., G.L. Foster, T.B. Chalk, et al. (2015) "Plio-Pleistocene Climate Sensitivity Evaluated Using High-Resolution CO2 Records." *Nature.* 518: 49–54.

Maslin, Mark A., Chris M. Brierley, et al. (2014) "East African Climate Pulses and Early Human Evolution." *Quaternary Science Reviews.* 101: 1–17.

Mayewski, P.A., L.D. Meeker, S. Whitlow, et al. (1993) "The Atmosphere During the Younger Dryas." *Science.* 261(5118): 195–197.

Mazrouei, Sara, Rebecca R. Ghent, William F. Bottke, et al. (2019) "Earth and Moon Impact Flux Increased at the End of the Paleozoic." *Science.* 363 (6424): 253–257.

McInerney, Francesca A. and Scott L. Wing. (2011) "The Paleocene-Eocene Thermal Maximum: A Perturbation of Carbon Cycle, Climate, and Biosphere with Implications for the Future." *Annual Review of Earth and Planetary Sciences.* 39: 489–516.

Mohd, Hamzah A., Jaffar A. Al-Tawfiq, and Ziad A. Memish. (2016) "Middle East Respiratory Syndrome Coronavirus (MERS-CoV) Origin and Animal Reservoir." *Virology Journal.* 13(87).

Monod, Jacques. (1969) "On Values in the Age of Science." In *The Place of Value in a World of Facts: Proceedings of the Fourteenth Nobel Symposium Stockholm, September 15–20, 1969.* Edited by Arne Tiselius and Sam Nilsson, 19–27. New York: Wiley Interscience Division.

Monod, Jacques. (1971) *Chance and Necessity.* Translated from the French by Austryn Wainhouse. New York: Alfred A. Knopf.

Muchhala, Nathan. (2006) "Nectar Bat Stows Huge Tongue in its Rib Cage." *Nature.* 444: 701–702.

Muchhala, Nathan and James D. Thomson. (2009) "Going to Great Lengths: Selection for Long Corolla Tubes in an Extremely Specialized Bat-Flower Mutualism." *Proceedings of the Royal Society B.* 276: 2147–2152.

Nei, Masatoshi. (2013) *Mutation-Driven Evolution*. Oxford: Oxford University Press.

Nordling, C.O. (1953) "A New Theory on Cancer-Inducing Mechanism." *British Journal of Cancer*. 7(1): 68–72.

Nowell, P.C. (1976) "The Clonal Evolution of Tumor Cell Populations." *Science*. 194(4260): 23–28.

O'Leary, Maureen A., Jonathan I. Bloch, John J. Flynn, et al. (2013) "The Placental Mammal Ancestor and the Post-K-Pg Radiation of Placentals." *Science*. 339(6120): 662–667.

Peacocke, Arthur. (1993) *Theology for a Scientific Age: Being and Becoming-Natural, Divine and Human*. Minneapolis: Fortress Press.

Pimiento, Catalina, John N. Griffin, Christopher F. Clements, et al. (2017) "The Pliocene Marine Megafauna Extinction and Its Impact on Functional Diversity." *Nature Ecology & Evolution*. 1: 1100–1106.

Pincock, Stephen. (2013) "Stephen Lyon Crohn." *The Lancet*. 382(9903): 1480.

Potts, Richard. (2013) "Hominin Evolution in Settings of Strong Environmental Variability." *Quaternary Science Reviews*. 73: 1–13.

Potts, Richard, Anna K. Behrensmeyer, J. Tyler Faith, et al. (2018) "Environmental Dynamics During the Onset of the Middle Stone Age in Eastern Africa." *Science*. 360: 86–90.

Rehan, Sandra M., Remko Leys, Michael P. Schwarz. (2013) "First Evidence for a Massive Extinction Event Affecting Bees Close to the K-T Boundary." *PLoS ONE*. 8(10): e76683.

Roach, Jared C., Gustavo Glusman, Arian F.A. Smit, et al. (2010) "Analysis of Genetic Inheritance in a Family Quartet by Whole-Genome Sequencing." *Science*. 328: 636–639.

Robertson, Douglas S., Malcolm C. McKenna, Owen B. Toon, et al. (2004) "Survival in the First Hours of the Cenozoic." *GSA Bulletin*. 116(5–6): 760–768.

Robertson, Douglas S., William M. Lewis, Peter M. Sheehan, and Owen B. Toon. (2013) "K-Pg Extinction: Reevaluation of the Heat-Fire Hypothesis." *Journal of Geophysical Research: Biogeosciences*. 118: 329–336.

Rodgers, Joseph Lee and Debby Doughty. (2001) "Does Having Boys or Girls Run in the Family?" *Chance*. 14(4): 8–13.

Rossiianov, Kirill. (2002) "Beyond Species: Il'ya Ivanov and His Experiments on Cross-Breeding Humans with Anthropoid Apes." *Science in Context*. 15(2): 277–316.

Rozhok and DeGregori (2019). "A Generalized Theory of Age-Dependent Carcinogenesis." *Cancer Biology*. eLife.39950.

Samet, Jonathan M., Erika Avila-Tang, Paolo Boffetta, et al. (2009) "Lung Cancer in Never Smokers: Clinical Epidemiology and Environmental Risk Factors." *Clinical Cancer Research*. 15(18): 5626–5645.

Sandin, S., D. Schendel, P. Magnusson, et al. (2016) "Autism Risk Associated with Parental Age and with Increasing Difference in Age Between the Parents." *Molecular Psychiatry*. 21: 693–700.

Schaller, Morgan F., Megan K. Fung, James D. Wright, et al. (2016) "Impact Ejecta at the Paleocene-Eocene Boundary." *Science*. 354(6309): 225–229.

Schaller, Morgan F. and Megan K. Fung. (2018) "The Extraterrestrial Impact Evidence at the Palaeocene-Eocene Boundary and Sequence of Environmental Change on the Continental Shelf." *Philosophical Transactions of the Royal Society A.* 376:20170081.

Schoffeniels, Ernest. (1976) *Anti-Chance: A Reply to Monod's Chance and Necessity.* Translated by B.L. Reid. Oxford: Pergamon Press.

Schulte, P., et al. (2010) "The Chicxulub Asteroid Impact and Mass Extinction at the Cretaceous-Paleogene Boundary." *Science.* 327: 1214–1218.

Secord, James A. (1981) "Nature's Fancy: Charles Darwin and the Breeding of Pigeons." *Isis.* 72(2): 162–186.

Shapiro, Michael D., Zev Kronenberg, Cai Li, et al. (2013) "Genomic Diversity and Evolution of the Head Crest in the Rock Pigeon." *Science.* 339(6123): 1063–1067.

Sharp, Paul M. and Beatrice H. Hahn. (2011) "Origins of HIV and the AIDS Pandemic." *Cold Spring Harbor Perspectives in Medicine.* 1: a006841.

Smit, J. (1999) "The Global Stratigraphy of the Cretaceous-Tertiary Boundary Impact Ejecta." *Annual Review of Earth Planet Sciences.* 27: 75–113.

Smit, J. and Hertogen, J. (1980) "An Extraterrestrial Event at the Cretaceous-Tertiary Boundary." *Nature.* 285: 198–200.

Snyder, Carolyn W. (2016) "Evolution of Global Temperature Over the Past Two Million Years." *Nature.* 538: 226–228.

Solloch, Ute V., Kathrin Lang, Vinznez Lange, et al. (2017) "Frequencies of Gene Variant Ccr5-Δ32 in 87 Countries Based on Next-Generation Sequencing of 1.3 Million Individuals Sampled from 3 National DKMS Donor Centers." *Human Immunology.* 78: 710–717.

Song, Huai-Dong, Chang-Chu Tu, Guo-Wei Zhang, et al. (2005) "Cross-Host Evolution of Severe Acute Respiratory Syndrome Coronavirus in Palm Civet and Human." *PNAS.* 102(7): 2430–2435.

Sproul, R.C. (1994) *Not a Chance: The Myth of Chance in Modern Science and Cosmology.* Grand Rapids, Michigan: Baker Academic.

Storz, Jay F. (2016) "Hemoglobin-Oxygen Affinity in High-Altitude Vertebrates: Is There Evidence for an Adaptive Trend?" *Journal of Experimental Biology.* 219: 3190–3203.

Sulloway, Frank J. (1982) "Darwin's Conversion: The Beagle Voyage and Its Aftermath." *Journal of the History of Biology.* 15(3): 325–396.

Tierney, Jessica E., Francesco S.R. Pausata, Peter B. deMenocal. (2017) "Rainfall Regimes of the Green Sahara." *Science Advances.* 3: e1601503.

Twain, Mark. (1935) *Mark Twain's Notebook.* New York: Harper & Bros.

USGS. (2015) "The Himalayas: Two Continents Collide." U.S. Geological Survey. pubs.usgs.gov/publications/text/himalaya.html

Vajda, Vivi and Antoine Bercovici. (2014) "The Global Vegetation Pattern Across the Cretaceous-Paleogene Mass Extinction Interval: A Template for Other Extinction Events." *Global and Planetary Change.* 122: 29–49.

Van Der Ryst, Elna. (2015) "Maraviroc—a CCR5 antagonist for the treatment of HIV-1 Infection." *Frontiers in Immunology.* 6(277): 1–4.

Van Wyhe, John. (2002). Editor. *The Complete Work of Charles Darwin Online.* www.darwin-online.org.uk.

Van Wyhe, John. (2007) "Mind the Gap: Did Darwin Avoid Publishing His Theory for Many Years?" *Notes and Records of the Royal Society.* 61: 177–205.

Varki, Ajit and Tasha K. Altheide. (2005) "Comparing the Human and Chimpanzee Genomes: Searching for Needles in a Haystack." *Genome Research.* 15: 1746–1758.

Vikrey, Anna I., Eric T. Domyan, Martin P. Horvath, and Michael D. Shapiro. (2015) "Convergent Evolution of Head Crests in Two Domesticated Columbids Is Associated with Different Missense Mutations in EphB2." *Molecular Biology and Evolution.* 32(10): 2657–2664.

Vogelstein, Bert, Nickolas Papadopoulos, Victor E. Velculescu, et al. (2013) "Cancer Genome Landscapes." *Science.* 339(6127): 1546–1558.

Vonnegut, Kurt. (1950) *Welcome to the Monkey House: A Collection of Short Works.* New York: Dial Press Trade Paperbacks.

Vonnegut, Kurt. (1998) *Timequake.* London: Vintage.

Vonnegut, Kurt. (1999) *God Bless You, Dr. Kevorkian.* Washington Square Press, Pocket Books: New York.

Vonnegut, Kurt. (2007) *A Man Without a Country.* New York: Random House.

Vonnegut, Kurt. (2009) *The Sirens of Titan.* New York: Dial Press Trade Paperbacks.

Vonnegut, Kurt. (2010) *Cat's Cradle.* New York: Dial Press Trade Paperbacks.

Vonnegut, Kurt. (2010) *Slapstick or Lonesome No More!* New York: Dial Press Trade Paperbacks.

Vonnegut, Kurt. (2014) *If This Isn't Nice, What Is?: Advice to the Young.* New York: Seven Stories Press.

Wain, Louise V., Elizabeth Bailes, Frederic Bibollet-Ruche, et al. (2007) "Adaptation of HIV-1 to Its Human Host." *Molecular Biology and Evolution.* 24(8): 1853–1860.

Walter, Christi A., Gabriel W. Intano, John R. McCarrey, C. Alex McMahan, and Ronald B. Walter. (1998) "Mutation Frequency Declines During Spermatogenesis in Young Mice but Increases in Old Mice." *Proceedings of the National Academy of Sciences.* 95: 10015–10019.

Ward, Keith. (1996) *God, Chance and Necessity.* Oxford: Oneworld Publications.

Wang, Jianbin, H. Christina Fan, Barry Behr, and Stephen R. Quake. (2012) "Genome-wide Single-Cell Analysis of Recombination Activity and De Novo Mutation Rates in Human Sperm." *Cell.* 150: 402–412.

Wang, Weina, Homme W. Hellinga, and Lorena S. Beese. (2011) "Structural Evidence for the Rare Tautomer Hypothesis of Spontaneous Mutagenesis." *Proceedings of the National Academy of Sciences.* 108(43): 17644–17648.

Watson, J.D. and F.H.C. Crick. (1953a) "Molecular Structure of Nucleic Acids: A Structure for Deoxyribose Nucleic Acid." *Nature.* 171(4356): 737–738.

Watson, J.D. and F.H.C. Crick. (1953b) "Genetical Implications of the Structure of Deoxyribonucleic Acid." *Nature.* 171(4361): 964–967.

Watson, J.D. and F.H.C. Crick. (1953c) "The Structure of DNA." *Cold Spring Harbor Symposia on Quantitative Biology.* 18: 123–131.

Watson, J.D. (1980) *The Double Helix: A Personal Account of the Discovery of the Structure of DNA.* New York: W.W. Norton & Company.

Watts, Caroline G., Martin Drummond, Chris Goumas, et al. (2018) "Sunscreen Use and Melanoma Risk Among Young Australian Adults." *JAMA Dermatology.* 154(9): 1001–1009.

Weart, Spencer. (2003) "The Discovery of Rapid Climate Change." *Physics Today.* 56(8): 30–36.

Wing, Scott L., Guy J. Harrington, Francesca A. Smith, et al. (2005) "Transient Floral Change and Rapid Global Warming at the Paleocene-Eocene Boundary." *Science.* 310: 993–996.

Zhang, Tao, Qunfu Wu, and Zhingang Zhang. (2020) "Probable Pangolin Origin of SARS-CoV-2 Associated with the COVID-19 Outbreak." *Current Biology.* 30: 1–6.

Zhu, Xiaojia, Yuyan Guan, Anthony V. Signore, et al. (2018) "Divergent and Parallel Routes of Biochemical Adaptation in High-Altitude Passerine Birds from the Qinghai-Tibet Plateau." *Proceedings of the National Academy of Sciences.* 115(8): 1865–1870.

Zhu, Yuan O., Mark L. Siegal, David W. Hall, and Dmitri A. Petrov. (2014) "Precise Estimates of Mutation Rate and Spectrum in Yeast." *Proceedings of the National Academy of Sciences.* 111: E2310–2318.

INDEX

Page numbers in *italics* refer to illustrations.